METHODS IN MOLECULAR BIOLOGY™

Series Editor
John M. Walker
School of Life Sciences
University of Hertfordshire
Hatfield, Hertfordshire, AL10 9AB, UK

For further volumes:
http://www.springer.com/series/7651

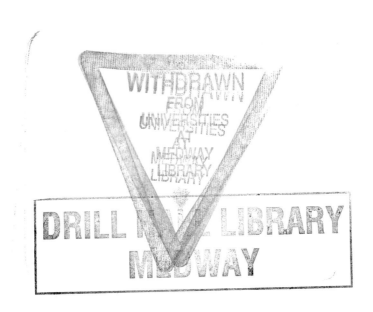

SELDI-TOF Mass Spectrometry

Methods and Protocols

Edited by

Charlotte H. Clarke

*Department of Translational Research, The University of Texas,
M.D. Anderson Cancer Center, Houston, TX, USA*

Diane L. Bankert McCarthy

Bio-Rad Laboratories, Malvern, PA, USA

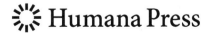

Editors
Charlotte H. Clarke
Department of Translational Research
The University of Texas
M.D. Anderson Cancer Center
Houston 77090, TX, USA
chclarke@mdanderson.org

Diane L. Bankert McCarthy
Bio-Rad Laboratories
Malvern, PA, USA
dbmccarthy68@verizon.net

ISSN 1064-3745   e-ISSN 1940-6029
ISBN 978-1-61779-417-9   e-ISBN 978-1-61779-418-6
DOI 10.1007/978-1-61779-418-6
Springer New York Dordrecht Heidelberg London

Library of Congress Control Number: 2011941205

© Springer Science+Business Media, LLC 2012
All rights reserved. This work may not be translated or copied in whole or in part without the written permission of the publisher (Humana Press, c/o Springer Science+Business Media, LLC, 233 Spring Street, New York, NY 10013, USA), except for brief excerpts in connection with reviews or scholarly analysis. Use in connection with any form of information storage and retrieval, electronic adaptation, computer software, or by similar or dissimilar methodology now known or hereafter developed is forbidden.
The use in this publication of trade names, trademarks, service marks, and similar terms, even if they are not identified as such, is not to be taken as an expression of opinion as to whether or not they are subject to proprietary rights.

Printed on acid-free paper

Humana Press is part of Springer Science+Business Media (www.springer.com)

Preface

This book provides an overview of the current applications of SELDI-TOF-MS (surface-enhanced laser desorption/ionization time-of-flight mass spectrometry), with an emphasis on study and experimental design, data analysis and interpretation, and assay development. SELDI is distinct from other time-of-flight mass spectrometer (TOF-MS) technologies in that it couples features of chromatography and mass spectrometry, facilitating analyte enrichment, and sample cleanup on an array surface. For the analyses of crude biological samples, all mass spectrometric techniques must eliminate substances found in the sample (i.e., buffer salts, detergents) which will interfere with MS detection and use fractionation or enrichment techniques to simplify the complex mixture. SELDI significantly reduces the need for off-line sample preparation by derivatized array surfaces to specifically capture and enrich proteins with specific biochemical properties. Washing of the arrays removes nonspecifically bound proteins and interfering compounds, greatly reducing signal suppression in the mass spectrometer caused by salts and detergents. For complex sample types, such as serum or plasma, SELDI analysis is often combined with upstream fractionation and enrichment techniques to maximize the number of unique proteins detected. After capture of the sample on the array, a matrix molecule, which enhances laser energy transfer and analyte ionization, is added to promote laser-based desorption and ionization. This part of the process is similar to MALDI-TOF-MS, but with a distinct advantage provided by the chemistry of the array. Proteins and peptides that have been enriched and retained on the array are then detected by the ProteinChip® SELDI System, a TOF-MS equipped with a laser desorption ion source.

In the growing field of proteomics, SELDI technology has been widely used for biomarker discovery and characterization in diverse applications including diagnostics, drug development, and basic research. Because chromatographic array surfaces capture proteins based on general chemistry rather than specific molecular affinity (as with antibody recognition), "undirected" studies can be performed to reveal novel biomarkers.

SELDI-based biomarker studies can typically be divided into four phases: discovery, validation, purification and identification, and assay development. In the discovery and validation phases, it is especially important to optimize study design and statistical methods to avoid preanalytical bias and yield robust markers. Identification can be performed at the end of the discovery phase or at any point during the validation phase, and generally requires standard protein purification procedures (column chromatography, size filtration, SDS-PAGE, etc.), followed by protease digestion and sequence analysis on a tandem mass spectrometer. Identification of the biomarkers provides insight into the disease biology and facilitates the development of analyte specific assays. Once the biomarkers have been positively identified, assays can be developed for routine testing on the most appropriate platform such as a quantitative immunoassay. SELDI-based immunoassays are particularly useful for detecting biomarkers with posttranslational modifications. This book provides information on optimizing study design, experimental protocols, and data analysis and interpretation to yield robust biomarkers and biomarker assays, using examples from different disease areas.

Fremont, CA, USA   *Charlotte H. Clarke*
Malvern, PA, USA   *Diane L. Bankert McCarthy*

Contents

Preface.. v
Contributors... ix

1 Optimized Conditions for a Quantitative SELDI TOF MS
  Protein Assay... 1
  *Lee Lomas, Charlotte H. Clarke, Vanitha Thulasiraman,
  and Eric Fung*

2 Solid-Phase Fractionation Strategies Applied
  to Proteomics Investigations.. 11
  Luc Guerrier, Frederic Fortis, and Egisto Boschetti

3 Data Processing and Analysis Using ProteinChip®
  Data Manager Software... 35
  Enrique A. Dalmasso and Dominic Caseñas

4 Purification and Identification of Candidate Biomarkers
  Discovered Using SELDI-TOF MS... 49
  Amanda L. Bulman and Enrique A. Dalmasso

5 Biomarker Discovery in Serum/Plasma Using Surface
  Enhanced Laser Desorption Ionization Time of Flight
  (SELDI–TOF) Mass Spectrometry... 67
  Momar Ndao

6 Plasma Proteomic Profiling of Pediatric Osteosarcoma.................... 81
  Yiting Li, Tu Anh Dang, and Tsz-Kwong Man

7 Profiling of Urine Using ProteinChip® Technology........................ 97
  *Ronald L. Woodbury, Diane L. Bankert McCarthy,
  and Amanda L. Bulman*

8 Protein Profiling of Cerebrospinal Fluid................................ 109
  Anja H. Simonsen

9 SELDI-TOF Mass Spectrometry-Based Protein Profiling
  of Tissue Samples for Toxicological Studies............................. 119
  Alexandra Sposny and Philip G. Hewitt

10 Proteomic Analysis of Skeletal Muscle Tissue Using
   SELDI-TOF MS: Application to Disuse Atrophy............................ 131
   Mark S.F. Clarke

11 Profiling Cervical Lavage Fluid by SELDI-TOF Mass Spectrometry......... 143
   Adam Burgener

12 Isolation and Proteomic Analysis of Platelets by SELDI-TOF MS.......... 153
   Sean R. Downing and Giannoula L. Klement

13 Using SELDI-TOF Mass Spectrometry on Amniotic Fluid
   and for Clinical Proteomics and Theranostics
   in Disorders of Pregnancy.. 171
   Irina A. Buhimschi

14 High Throughput Profiling of Serum Phosphoproteins/Peptides
   Using the SELDI-TOF-MS Platform 199
   Lin Ji, Gitanjali Jayachandran, and Jack A. Roth

15 Analysis of Protein–Protein Interaction Using ProteinChip
   Array-Based SELDI-TOF Mass Spectrometry 217
   Gitanjali Jayachandran, Jack A. Roth, and Lin Ji

16 Quantitation of Amyloid Beta Peptides in CSF by Surface
   Enhanced MALDI-TOF 227
   *Eddie Takahashi, Anita Howe, Ole Vesterqvist,
   and Zhaosheng Lin*

Index *237*

Contributors

EGISTO BOSCHETTI • *Bio-Rad Laboratories, Marnes la Coquette, France*
IRINA A. BUHIMSCHI • *Department of Obstetrics, Gynecology and Reproductive Sciences, Yale University School of Medicine, New Haven, CT, USA*
AMANDA L. BULMAN • *Bio-Rad Laboratories, Malvern, PA, USA*
ADAM BURGENER • *National Laboratory for HIV Immunology, Public Health Agency of Canada, Winnipeg, MB, Canada*
DOMINIC CASEÑAS • *Bio-Rad Laboratories, Hercules, CA, USA*
CHARLOTTE H. CLARKE • *Department of Translational Research, The University of Texas, M.D. Anderson Cancer Center, Houston, TX, USA*
MARK S.F. CLARKE • *Department of Health and Human Performance, University of Houston, Houston, TX, USA*
ENRIQUE A. DALMASSO • *Bio-Rad Laboratories, Hercules, CA, USA*
TU ANH DANG • *Cell Biosciences, Inc., Santa Clara, CA, USA*
SEAN R. DOWNING • *Vascular Biology Program, Children's Hospital Boston, Boston, MA, USA*
FREDERIC FORTIS • *Bio-Rad Laboratories, Marnes la Coquette, France*
ERIC FUNG • *Vermillion, Inc., Fremont, CA, USA*
LUC GUERRIER • *Bio-Rad Laboratories, Marnes la Coquette, France*
PHILIP G. HEWITT • *Merck Serono Research, Merck KGaA, Darmstadt, Germany*
ANITA HOWE • *Wyeth Research, Collegeville, PA, USA*
GITANJALI JAYACHANDRAN • *Department of Thoracic and Cardiovascular Surgery, The University of Texas, Anderson Cancer Center, Houston, TX, USA*
LIN JI • *Department of Thoracic and Cardiovascular Surgery, The University of Texas, Anderson Cancer Center, Houston, TX, USA*
GIANNOULA L. KLEMENT • *Children's Hospital, Boston, MA, USA*
YITING LI • *Texas Children's Cancer Center and Department of Pediatrics, Baylor College of Medicine, Houston, TX, USA*
ZHAOSHENG LIN • *Wyeth Research, Collegeville, PA, USA*
LEE LOMAS • *Laboratory Separations Division, Bio-Rad Laboratories, Hercules, CA, USA*
TSZ-KWONG MAN • *Texas Children's Cancer Center and Department of Pediatrics, Baylor College of Medicine, Houston, TX, USA*
DIANE L. BANKERT MCCARTHY • *Bio-Rad Laboratories, Malvern, PA, USA*
MOMAR NDAO • *National Reference Centre for Parasitology, Research Institute of the McGill University Health Centre, Montreal General Hospital, Montreal, QC, Canada*
JACK A. ROTH • *Department of Thoracic and Cardiovascular Surgery, The University of Texas, Anderson Cancer Center, Houston, TX, USA*
ANJA H. SIMONSEN • *Department of Neurology, Copenhagen University Hospital, Rigshospitalet, Copenhagen, Denmark*

ALEXANDRA SPOSNY • *Merck Serono Research, Merck KGaA, Darmstadt, Germany*
EDDIE TAKAHASHI • *Wyeth Research, Collegeville, PA, USA*
VANITHA THULASIRAMAN • *Laboratory Separations Division, Bio-Rad Laboratories, Hercules, CA, USA*
OLE VESTERQVIST • *Wyeth Research, Collegeville, PA, USA*
RONALD L. WOODBURY • *Bio-Rad Laboratories, Malvern, PA, USA*

Chapter 1

Optimized Conditions for a Quantitative SELDI TOF MS Protein Assay

Lee Lomas, Charlotte H. Clarke, Vanitha Thulasiraman, and Eric Fung

Abstract

The development of peptide/protein analyte assays for the purpose of diagnostic tests is driven by multiple factors, including sample availability, required throughput, and quantitative reproducibility. Laser Desorption/ionization mass spectrometry methods (LDI-MS) are particularly well suited for both peptide and protein characterization, and combining chromatographic surfaces directly onto the MS probe in the form of surface enhanced laser desorption/ionization (SELDI)-biochips has improved the reproducibility of analyte detection and provided effective relative quantitation. Here, we provide methods for developing reproducible SELDI-based assays by providing a complex artificial protein matrix background within the sample to be analyzed that allows for a common and reproducible ionization background as well as internal normalization standards. Using this approach, quantitative assays can be developed with CVs typically less than 10% across assays and days. Although the method has been extensively and successfully implemented in association with a protein matrix from *E. coli*, any other source for the complex protein matrix can be considered as long as it adheres to a set of conditions including the following: (1) the protein matrix must not provide interferences with the analyte to be detected, (2) the protein matrix must be sufficiently complex such that a majority of ion current generated from the desorption of the sample comes from the complex protein matrix, and (3) specific and well-resolved protein matrix peaks must be present within the mass range of the analyte of interest for appropriate normalization.

Key words: SELDI-MS, Quantitation, Transthyretin, Ovarian cancer

1. Introduction

The objective of biomedical research is to provide adequate answers to biological questions that impact human health. This ranges from detection of diseases at continually earlier time points to better understanding the progression of disease and finally developing suitable intervention methods for better patient care and management. Whether the question is asked at the earliest research level or at the clinic, the primary goal is to define a routine and consistent method to generate the appropriate answer; typically in the form of an assay.

Analytical assays typically fall into two general categories: those coupled to indirect detection and those coupled to direct detection. Indirect detection methods, such as ELISA, RIA, Nephelometry, or enzymatic amplification are well established and once correctly developed offer highly robust and reproducible assays with high throughput and low costs (1). Direct detection methods based on mass spectrometry have also become well established and techniques such as gas or liquid chromatography-mass spectrometry (GC/LC-MS^n, where n represents tandem processes) are commonplace in analytical diagnostic settings, particularly for the quantitation of small molecules and peptides (2–4). One advantage of MS methods over indirect detection methods is that they offer information regarding analyte conformation due to specific mass and/or fragmentation information that is provided by the mass analyzer.

In recent years, a greatly expanded effort has begun to focus on proteomics and its diagnostic applications. The term "biomarker" is now almost ubiquitous throughout the proteomic literature, and there is considerable effort in correlating such indicators with disease events. Time of flight (TOF) mass spectrometry (MS) is now a common method of peptide and protein characterization and can detect singly charged proteins of high mass (greater than 200 kDa) in a high throughput mode. A specific type of TOF-MS known as surface enhanced laser ionization/desorption mass spectrometry (SELDI-MS) allows for the capture, or retention, of protein molecules on the surface of chemically modified arrays that are subsequently detected in a time of flight mass spectrometer (5) (Fig. 1). The chemical modifications of the arrays that are commercially available (Bio-Rad) provide ionic, metal-binding, or hydrophobic surfaces (Fig. 1a). The fundamental principle behind time of flight mass spectrometry is that smaller and more highly charged molecules arrive at the detector sooner than larger molecules (Fig. 1b). SELDI data are presented in a spectrogram in which the y-axis represents intensity (essentially the number of particles hitting the detector at a given time) and the x-axis represents mass (more precisely, the mass to charge ratio), which is derived from an equation relating time of flight to mass (Fig. 1c). Protein peaks are designated as intensity signals that exceed a specific criterion (e.g., 3× the signal to noise ratio), and the mass of the protein peak corresponds to the x-axis value where the y-axis achieves that threshold. A typical spectrogram taken from fractionated human serum can have 100–200 protein peaks.

This chapter presents an approach that overcomes many of the difficulties associated with the use of TOF-MS for a quantitative assay, particularly in situations where high levels of heterogeneity are observed between samples and calibrants. The method presented here also allows for normalization across MS instruments

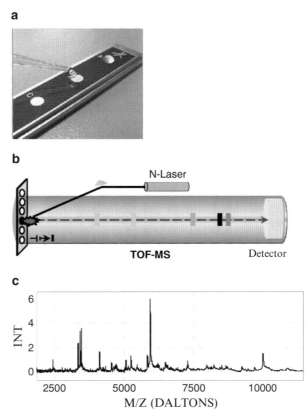

Fig. 1. (a) spots on a ProteinChip array (b) schematic of laser desporption/ionization time of flight mass spectrometry (c) example of a TOF-mass spectrum.

and locations to provide a completely portable assay. One protein marker currently being measured in a SELDI-based assay under development as a marker of ovarian cancer is transthyretin (TT), which is present in serum in multiple isoforms. The quantity of TT measured in serum of normal individuals or patients with benign disease falls within a specific range, typically 15–36 mg/dL. Likewise, measurements of these proteins, which are outside of the specified range (usually below), indicate the possibility of malignant disease in that patient.

2. Materials

2.1. Mass Spectrometry and Related Consumables

1. ProteinChip System SELDI mass spectrometer (Bio-Rad).
2. Q10 ProteinChip Arrays (strong anionic exchangers, Bio-Rad).
3. Sinapinic Acid, saturated solution in 50% acetonitrile (VWR) and 0.5% aqueous trifluoroacetic acid (Pierce).
4. SELDI array bioprocessor (Bio-Rad).

2.2. General Reagents

1. Phosphate-buffered saline (0.1 M, pH 7).
2. *Escherichia coli* extract (Paragon Biosciences, 0.1 mg/mL in 0.1 M PBS, pH 7).
3. 96-well microtiter plates (VWR).

2.3. Biological Calibrators

1. Reference Intergen human serum (Serologic Inc, pooled normal human serum).

3. Methods

3.1. Generation of a Transthyretin (Cysteinylated Isoform) Quantitative Assay

TT represents a high abundance protein within serum that has strong binding affinity to strong anion exchange ProteinChip arrays (Q10 Arrays). Owing to the absence of a commercial source of TT-depleted serum with which one can resupplement with known amounts of TT, the simplest method of generating a standard curve is to quantify the endogenous amount of TT in reference serum by an independent method, such as ELISA or Nephelometry, then preparing a serial dilution of this reference serum to generate respectively decreasing TT amounts. MS spectra from such a serial dilution of reference serum (Intergen) are presented in Fig. 2a. As expected, TT (both the unmodified and cysteinylated forms at 13,764.2 Da and 13,880.2 Da respectively) are efficiently extracted from the serum sample on the anionic

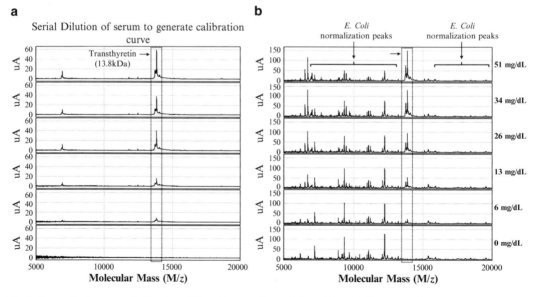

Fig. 2. The method of normalization by addition of a complex protein matrix. (a) MS spectra of a decreasing calibration series of cysteinylated transthyretin (*from top to bottom*) generated from serial dilution of serum into buffer only or (b) into buffer containing a constant amount of *E. coli* extract.

exchanger and represent the dominant ion signal pair throughout the spectra. Under this situation, the normalization methods commonly used in research protocols that are based on total ion current measurement would artificially compress the calibration curve and not be appropriate. However, when these serum samples are supplemented with a complex protein mixture such as *E. coli* extract (Note 1–3), the overall protein profile now generated by MS is predominantly similar across all calibrants due to the presence of the retained and constant background of *E. coli* peaks (Fig. 2b). Although the method has been successfully and extensively implemented in association with a protein matrix derived from *E. coli*, any source for the complex protein matrix can be considered as long as it adheres to a set of specific conditions including the following: (1) the protein matrix must not provide interferences with the analyte to be detected, (2) proteins derived from the complex matrix must represent a substantial component of the total proteins (including the analyte to be quantitated) presented to the ion source of the MS such that a majority of ion current generated from the desorption of the sample comes from these complex matrix proteins, and (3) specific and well-resolved protein matrix peaks must be present within the mass range of the analyte of interest for appropriate normalization.

3.2. Method of Assay Normalization

E. coli peaks represent both an internal mass and quantity standard that can be subsequently used to internally calibrate the MS, as well as provide a method of ion current normalization to help compensate for variability in the assay manipulation and ionization within the MS ion source. To accomplish this, 10 *E. coli* proteins with the following characteristics are selected: (1) all *E. coli* peaks are clearly resolved from any endogenous serum protein peaks to allow for consistent and unencumbered peak selection and ion current measurement; (2) all *E. coli* peaks have signal-to-noise ratios of at least 30; and (3) all *E. coli* peaks are within 10 kDa of the target analyte peak both above and below the molecular mass of the target analyte. Figure 3 contains representative spectra of the 51 mg/dL TT concentration without added *E. coli* (Fig. 3a), the 51 mg/dL TT concentration with added *E. coli* (Fig. 3b), and the 0 mg/dL TT concentration with added *E. coli* (*E. coli* extract without serum (Fig. 3c)). Figure 3c also highlights the selected *E. coli* peaks with their masses labeled.

3.3. Confirmation of the Absence of E. coli Protein Matrix Interferences

Figure 3b, c also shows a direct comparison (right panels) for the determination of the presence of any interfering proteins within the critical TT mass range, as well as the selection of unique *E. coli* peaks that can later be used for the normalization process. In this case, no interfering peaks are observed within the critical TT mass range and a selection of ten peaks (five below and five above the mass of TT) that are further used for the normalization process.

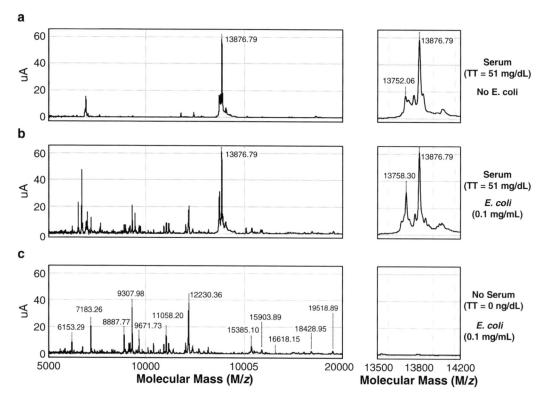

Fig. 3. Evaluation of interferences and selection of normalization peaks. (**a**) represents serum alone, while (**b**) represents serum spiked with 0.1 mg/mL *E. coli* extract. (**c**) represents *E. coli* only (0 mg/mL TT) and is also used to assess interference.

3.4. Preparation of Standard Curves for the Measurement of TT

Standard curves in triplicate are generated by preparing a serial dilution of reference Intergen serum in TT sample buffer (0.1 M PBS, pH 7 + *E. coli* lysate at 0.1 mg/mL). Calibrants are prepared to include TT reference concentrations at 51.4 mg/dL, 34.3 mg/dL, 25.5 mg/dL, 12.8 mg/dL, 6.4 mg/dL, and 0 mg/dL and measured in 50 µL triplicates.

3.5. Preparation of Unknown Samples for the Measurement of Transthyretin

Serum samples are thawed and 5 µL diluted into 1,245 µL TT sample buffer. Three replicates of 50 µL per replicate are randomized on a 96-well microtiter footprint for quantitation.

3.6. Preparation of Reference Samples for the Measurement of TT

Reference Intergen human serum (Serologic Inc.) aliquot is thawed at the time of assay and 24 µL serum diluted into 5,976 µL TT sample buffer. 12 replicates of 50 µL per replicate reference serum are randomized on a 96-well microtiter footprint for quantitation. The reference Intergen sample contains 25.5 mg/dL total TT as determined by ELISA.

3.7. SELDI Transthyretin Assay

12 Strong anionic ProteinChip® Arrays (Q10, Bio-Rad) were assembled into a 96-well bioprocessor cassette with associated bioprocessor top (Bio-Rad) and equilibrated with 150 µL of conditioning buffer (0.1 M PBS, pH 7) for 30 min. The conditioning buffer is removed and standard curves, reference Intergen serum and unknown samples are distributed across the arrays. Samples are incubated on the array spots for 2 h at room temperature, and then removed. Array spots are subsequently washed three times with 150 µL conditioning buffer, followed by two washes of 150 µL water per wash. Water is removed and array spots allowed to air-dry. For all steps, sample application and array washing should be accomplished using robotic liquid handling equipment such as the TECAN Aquarius robotic workstation. After array air-dry, 0.75 µL sinapinic acid (50% MeCN, 0.5% aqueous TFA) is added (twice with drying between) using a BioDot robotic workstation (Note 4). Finally, arrays are loaded into a PCS4000 ProteinChip System (Bio-Rad) for data collection.

3.8. Data Processing

After peak labeling the peak intensity values are exported to Excel and a normalization calculation performed based on analyte peak intensity/average (*E. coli* peak intensity). This normalized analyte intensity is then plotted against the corresponding known concentration of analyte to generate a suitable standard curve. Figure 4 and Table 1 show this curve (generated as an average of triplicates), as well as the calculated concentration of the reference Intergen serum based on this curve. The calculated concentration of TT in Intergen (25.8 mg/dL) agrees well with that measured independently by ELISA (27.5 mg/dL). Additionally, the % CV of this measurement is below 11%.

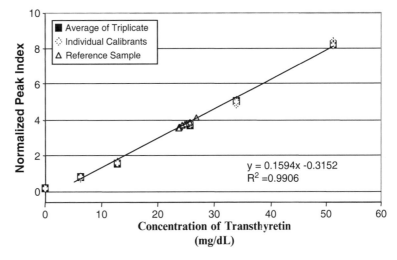

Fig. 4. Generation of transthyretin standard curves and quantitation of a reference serum sample.

Table 1
Measured concentration of transthyretin as measured in a reference serum sample

	N	Calculated conc. of TT in Intergen serum (mg/dL)	CV%
Plate 1	15	25.90	6.80
Plate 2	12	24.38	3.92
Plate 3	17	27.04	6.08
Plate 4	12	25.91	5.06
Plate 5	15	26.10	5.82
Plate 6	15	21.97	8.82
Plate 7	15	29.40	6.54
Plate 8	15	26.89	6.62
Plate 9	15	25.97	8.07
Plate 10	15	23.78	10.51

The measured transthyretin concentration (25.8 mg/dL) was in accordance with the known transthyretin concentration (25.5 mg/dL) as determined by ELISA
CVs were typically less than 10%

3.9. Application

The development of MS-based quantitative assays for protein analytes typically relies on the availability of certain reagents that allow for the generation of a suitable standard curve and suitable normalization. Generation of differential standards based on stable isotopes are well suited for peptides and typically used in electrospray – MS; however, such standards are difficult and expensive to prepare for larger peptides and proteins. Additionally, they require specific preparation and subsequent purification that in themselves can be problematic.

The described method provides an approach that solves many fundamental problems associated with TOF-MS and reproducibility. This includes providing sufficient control over the chromatographic reproducibility when operated in a "batch" separation mode and irreproducibility that occurs during the ionization phase of the analysis due to different sample compositions. The example provided here demonstrates the possible improvement that can be expected. A quantitative MS assay for a protein such as TT can be problematic, as it is an abundant protein and binds strongly to a strong anionic exchanger (Q10 ProteinChip Array). In batch binding mode, conditions are such that most other proteins are excluded from surface binding and thus the TT isoforms almost exclusively dominate the spectra (Fig. 2a). Additionally, TT depleted serum is not commercially available and the most suitable method to generate the standard titration curve is by serial dilution of the serum into the binding buffer. This defines a situation where the overall composition of the lowest calibrant, the highest calibrant, and the unknown sample can be dramatically different, and affecting the binding characteristics during the chromatography. Finally, because the TT dominates the MS spectra, normalization methods based on total ion current are not effective.

The addition of *E. coli* as a complex protein matrix provides a unique situation whereby all these situations can be addressed. First, *E. coli* can be added in sufficient quantity to effectively compete with the dominant TT isoforms; this is demonstrated by the MS spectra of Fig. 2b. As the amount of E. coli extract added to each sample is held constant, a standard curve of TT can be generated by serially diluting serum into this protein matrix with a minimum impact on the chromatographic retention of TT from calibrants or unknown samples.

The complex pattern of proteins presented to the ionization source in the mass spectrometer also is of benefit during the MS detection step as well. SELDI MS requires the protein sample to be cocrystallized with an organic compound such as α-cyano-4-hydroxycinnamic acid (CHCA; typically used for peptides and small proteins below 10 kDa) or sinapinic acid (SPA; typically used for proteins above 10 kDa) that promotes codesorption and analyte ionization upon laser irradiation; a process that is not well understood. However, to generate the greatest reproducibility for this process it is best to hold constant as many parameters as possible. By providing a complex protein matrix to the ionization source that is as similar as possible across all samples greatly enhances a consistent ionization event and removes variability.

Finally, because a constant amount of *E. coli* extract is added to all samples, the individual *E. coli* peaks can be selected and used as a final normalization method across all samples. Although a total of ten *E. coli* peaks are chosen in this example (five above and five below the mass of our TT analyte) the selection of as few as three or four can provide a similar improvement.

The exploration of this approach may also have interesting implications on how we consider the chromatographic capture of the analyte of interest. This work with the TT assay suggests that when a sample is supplemented with proteins that have an equal or higher affinity as compared to the analyte, a situation of displacement chromatography can be considered. In such a mode, the overall capacity of the surface can be better utilized based on more efficient binding conditions.

In this method, *E. coli* is used as the source of the complex protein matrix and detection is restricted to SELDI. However, it is also clear that any complex protein mixture could potentially be used and the choice may ultimately be dictated by the analyte to quantitate and chromatographic surface used. How "complex" the matrix must be remains to be determined; however, preliminary experiments using much simpler mixtures, for example mixes of five proteins, provide inferior results. This approach may have an equal benefit to a quantitative matrix assisted laser desorption/ionization (MALDI) TOF-MS application where the up-front chromatographic step is carried out in batch mode such as when chromatographic resins are used in a microtiter filter plate followed by deposition of eluted proteins onto a MALDI target.

4. Notes

1. The protein matrix chosen, in this case *E. coli*, must not provide interferences with the analyte to be detected.
2. Proteins derived from the complex matrix must represent a substantial component of the total proteins (including the analyte for quantitation) presented to the ion source of the MS such that a majority of ion current generated from the desorption of the sample comes from these complex matrix proteins.
3. Specific and well resolved protein matrix peaks must be present within the mass range of the analyte of interest for appropriate normalization.
4. Drying times and conditions for matrix deposition can contribute significantly to variability. Drying times for all arrays, as well as environmental temperature and humidity, should be as consistent as possible.

References

1. Dhawan, S. (2006) Signal amplification systems in immunoassays: implications for clinical diagnostics. *Expert Review of Molecular Diagnostics*, 6:749–760.
2. Wudy, S.A. and Hartmann, M.F. (2004) Gas chromatography-mass spectrometry profiling of steroids in times of molecular biology. *Hormone and Metabolic Research*, 36:415–22.
3. Ma, S., Chowdhury, S.K. and Alton, K.B. (2006) Application of mass spectrometry for metabolite identification. *Current Drug Metabolism*, 7:503–23.
4. Carpenter, KH and Wiley, V. (2002) Application of tandem mass spectrometry to biochemical genetics and newborn screening. *Clinical Chimica Acta*, 322:1–10.
5. Weinberger, S.R.,. Lomas, L.O. Fung, E.T. and Enderwick, C. (2007) Surface Enhanced Laser Desorption/Ionization Protein Biochip Technology for Proteomics Research and Assay Development. In: *Spectral Techniques in Proteomics* (ed D. Sem). Taylor & Francis Group, LLC. 101–132.

Chapter 2

Solid-Phase Fractionation Strategies Applied to Proteomics Investigations

Luc Guerrier, Frederic Fortis, and Egisto Boschetti

Abstract

Methods for protein fractionation in the proteomics investigation field are relatively numerous. They apply to the prefractionation of the sample to obtain less complex protein mixtures for an easier analysis; they are also used as a means to evidence specific proteins or protein classes otherwise impossible to detect. They involve depletion of high-abundance proteins suppressing the signal of dilute species; they are also capable to enhance the detectability of low-abundance species while concomitantly decreasing the concentration of abundant proteins such as albumin in serum and hemoglobin in red blood cell lysates. Fractionation of proteomes is also used for the isolation of targeted species that are selected for their different expression under certain pathological conditions and that are detected by mass spectrometry. Two unconventional methods of large interest in proteomics due to the low level of protein redundancy between fractions are also reported.

All these methods are reviewed and detailed method given to allow specialists of proteomics investigation to access selected separation methods generally dispersed on different technical reviews or books.

Key words: Protein fractionation, Chromatography, Depletion, Concentration range, Identification, Proteomics

1. Introduction

The very large complexity of proteomes in terms of protein diversity and concentration differences is not any longer to be demonstrated. If we limit the analysis to human proteome, already in 2002, Anderson underlined the very large diversity and an extremely wide dynamic concentration range (1). With the expression of a large number of genes, splicing proteins and stitching different portions together, then cutting out amino acids sequences from the melted pieces and making a number of posttranslational modifications, the capability of living cells to make protein species is extremely large.

In this very intricate situation, it is clear that an obvious way to analyze the content of a proteome is the fractionation into discrete groups for individual analysis. A difficulty that is encountered is the extremely large difference in concentration of individual proteins that can range up to 12 and even more orders of magnitude (1) with extreme examples where one protein dominates very largely, such as hemoglobin in red blood cells representing about 98.5% of total protein mass. As a consequence of such differences in protein concentration, fractionation represents a hard task. This is the reason why sample pretreatments such as removal of proteins of high abundance (2–4) prior fractionation or reduction of the dynamic concentration range (5–8) are proposed methods that encounter a great success.

Fractionation gives access to many more proteins, including some dilute ones as a consequence of the concentration effect due to solid-phase adsorption. A number of reviews have been published on the theme of protein prefractionation using a variety of approaches; a recent presentation of prefractionation methods has been made by Herbert et al. (9). Prefractionation tools comprise not only all possible chromatographic and electrophoretic methods (especially those based on isoelectric focusing) but also organelle prefractionation (10–12).

In spite of numerous investigational approaches and hundreds of published reports, major limitations still exist to getting access to whole proteomes. Highly hydrophobic proteins such as membrane proteins cannot be properly solubilized and consequently are not easily analyzed and/or identified. Strongly alkaline proteins are also poorly represented using classical two-dimensional electrophoresis (13). Moreover, posttranslational modifications and especially glycosylations are still very difficult to analyze.

The centrifugal cell-fractionation method is well-ingrained since its initial development (14). Sub-cellular organelles are isolated in reasonably pure form such as mitochondria, nuclei, lysosomes, microbodies, synaptosomes, and the like. This is the most direct and efficient method for enrichment of the desired protein fractions as they are part of specific substructures of the cells. This fractionation technique is, however, challenged by other techniques such as electrical and solid-phase based separations methods.

Preparative fractionation methods based on differential migration under an electrical field cover a wide range of technologies, the most popular being free-flow fractionation (15), multicompartment isoelectric focusing (16, 17) and the so-called Off-Gel fractionation (18). While electrophoretic methodologies, in all of their variegated aspects, may represent a high potential especially for their efficiency, they are dependent on the use of samples with preestablished low ionic strength and, for pI-based separation, they require an added carrier ampholine that is problematic for further mass spectrometry analytical determinations.

This document focuses in particular on solid-phase-based separation techniques involving in particular beaded material such as a variety of chromatographic fractionation methods. Various types of protein chromatography are currently used in proteome investigation; they are based on ion-exchange effect (19, 20), (cation and anion exchange), hydrophobic interactions (21), chromatofocusing (22), molecular sieving (23), and a variety of affinity or affinity-like interactions. Chromatographic fractionation gives access to many more proteins, including some dilute ones as a consequence of the concentration effect due to solid-phase adsorption. A number of reviews have been published on the theme of protein prefractionation using a variety of chromatographic approaches (24–28); an extensive description of prefractionation methods has been made by Herbert et al. (9) and more recently by Jmeian et al. (29). However, chromatographic approaches much in vogue today regard the capture of classes of proteomes based on their structure or function, such as the specific seizure of the glycoproteome via lectin columns (30–33) or the capture of the phosphoproteome via chelated metal ions (Fe^{3+}, Ga^{3+}, Ti^{4+}, Zr^{4+}) on solid phases (34).

2. Materials

2.1. Sample Pretreatment

2.1.1. High-Abundance Protein Depletion

1. Immunosorbents capable of removing 6 or 12 different high-abundance proteins from human serum are available from various vendors (e.g., Agilent, Palo Alto, CA, Sigma-Aldrich, Saint Louis, MO or Beckman Coulter Fullerton CA) with the capacity to treat 20–25 µL of human serum per run.
2. Current buffers used in biochemistry are needed, including the following:
   (a) 10 mM Tris–HCl, 150 mM sodium chloride, pH 7.4 for dilution and washing.
   (b) 100 mM glycine–HCl, pH 2.5 for stripping.
   (c) 100 mM Tris–HCl, pH 8 for neutralization.

2.1.2. Reduction of Dynamic Protein Concentration Range

Three main ingredients are necessary for this approach. Elution can be performed as a single step or in multiple steps to further fractionate the eluate.

1. The solid-phase library (ProteoMiner beads, Bio-Rad, Hercules, CA), a washing buffer (phosphate-buffered saline, PBS).
2. Elution buffers
   (a) For single-step elution: 9 M urea containing 5% acetic acid and 2% CHAPS.

(b) For multistep elution, several elution buffers are used sequentially. Two main options are available:
- Option 1:
  – 1 M sodium chloride in 20 mM Hepes.
  – 200 mM glycine–HCl pH 2.4.
  – 60% ethylene glycol in water.
  – Hydro-organic mixture composed of 33.3% isopropanol, 16.7% acetonitrile and 0.1% TFA in DI water.
- Option 2:
  – 1 M sodium chloride.
  – 7 M urea, 2 M thiourea, 2% CHAPS.
  – 9 M urea containing 50 mM citric acid and 2% CHAPS.

3. Protease inhibitor cocktail (Sigma-Aldrich, Saint Louis, MO).

2.1.3. Classical Chromatography Separation Using Fractionated Elution

1. Solid-phase media such as ion exchangers, gel-filtration media, hydrophobic resins and affinity-based sorbents can be obtained from current manufacturers of chromatographic materials such as Bio-Rad Laboratories (Hercules, CA), General Electric (Uppsala, Sweden), or Biosepra (Cergy, France).
2. Chromatographic columns of various dimensions are obtained from solid-phase media suppliers.
3. Buffers used for protein adsorption are made of Tris–HCl, sodium acetate–acetic acid, phosphates, or "good" buffers. The pH and ionic strength are dependent from the type of chromatographic separations:
   (a) For cation exchange buffer, pH is between 4 and 6 and ionic strength is relatively low (10–25 mM).
   (b) For anion exchange buffer, pH is between 7.5 and 9 and ionic strength is relatively low (10–25 mM).
   (c) For gel filtration, the buffer is generally neutral and ionic strength corresponding to 100 mM sodium chloride.
   (d) For hydrophobic chromatography, high concentrations of lyotropic salts (such as 0.8 M ammonium sulfate) are added to the adsorption buffer at neutral pH.
   (e) Affinity-based separations are most generally started with physiological buffers.
4. Protein elution (except for gel filtration that is an isocratic chromatography) is obtained by:
   (a) For ion exchangers: fractionated desorption by increasing the ionic strength of buffer.
   (b) For hydrophobic separations: decreasing the concentration of lyotropic salts.

(c) For affinity chromatography, elute using either displacements with specific or nonspecific agents or by simply using acidic pH (200 mM glycine–HCl, pH 2.4 or 100–200 mM acetic acid).

2.1.4. Fractionation with Stacked Sorbents and Single Buffer

1. Chromatography media can be obtained from specialized manufacturers such as Bio-Rad Laboratories (Hercules, CA), General Electric (Uppsala, Sweden), Biosepra, (Cergy, France), or Sigma-Aldrich (Saint Louis, MO).
2. Columns are selected from minicolumns that can be superimposed or any other practical device having systems connectable each other such as luer-lock capabilities between the upper column and the lower column.
3. The first buffer used is generally a Tris–HCl buffer. Elution buffer is any kind of stripping solution that is compatible with downstream analytical methods. Good eluting agents include the following:
   (a) 9 M urea containing 2% CHAPS and 2% ammonium hydroxide or
   (b) A mixture composed of 20% ammonium hydroxide (8 volumes)–distilled water (72 volumes)–acetonitrile (6.6 volumes) and isopropanol (13.4 volumes).

2.2. pI Separation Groups by Solid-Phase Buffers

1. Solid-State Buffers are made as described by Fortis et al. (35).
2. Ion exchangers are selected from those proposed by current producers such as Bio-Rad Laboratories (Hercules CA), General Electric (Uppsala, Sweden), or Biosepra (Cergy, France).
3. Columns are selected from minicolumns that can be superimposed or any other practical device having systems connectable each other such as luer-lock capabilities between the upper column and the lower column. Small fixed columns are from Promega (Madison, WI) or Omnifit (Cambridge, UK).
4. Urea.
5. Potassium chloride.
6. Sodium chloride.
7. Tris (hydroxymethyl) aminomethane.
8. Sodium acetate.
9. Acetic acid.

3. Methods

3.1. Sample Pretreatment

Since it is unlikely that improved sensitivity instruments and methods could go deep enough to detect trace proteins, the pretreatment of biological samples represents a possible way to improve detectability of very dilute gene products. At present, it is unanimously admitted that the detectability of trace proteins is mostly hindered by both the presence of high-abundance proteins and the too low concentration of a number of species. To resolve this issue, several approaches have been proposed, but two of them are dominating:

- Removing high-abundance species that hide minor species, such as via so-called depletion approaches.
- Reducing the dynamic concentration range by reducing the concentration of high-abundance species while concomitantly and proportionally concentrating low-abundance proteins.

Detailed descriptions of these two methods are given below. These methods could be either used singularly or in association.

3.1.1. High-Abundance Protein Depletion Methods

The term "depletion" is used to indicate the selective removal of one or several proteins from a complex mixture by means of a selective solid-phase media as an immunoadsorbent. Most generally, proteins that are removed include albumin, immunoglobulins, and transferrin from serum or plasma (36), hemoglobin from red blood cell lysates (37), actin from cell extracts (38), and ribulose-bisphosphate-carboxylase-oxygenase from leaf extracts (39). The removal of these high-abundance proteins allows evidencing the presence of other species by two-dimensional electrophoresis as well as by mass spectrometry due to elimination of suppression effects in both cases.

Various depletion strategies have been proposed over the past few years involving various sorbents for the removal of undesired proteins and to analyze the resulting depleted sample directly by 2-DE and mass spectrometry (40). Dye ligands, or other natural proteinaceous ligands such as protein A and G, lectins, and synthetic ligands are among described techniques. Nevertheless, the most used methods involve antibodies against proteins targeted for removal.

In spite of better visibility of hidden proteins, specific drawbacks limit the use of depletion methods:

- Codepletion (total or partial) of associated species due to non-specific binding or to protein–protein interaction (41).
- Limited binding capacity of immunosorbents with consequently very small volumes of biological extract treated.
- Dilution of the initial extract rendering it even more difficult to detect already diluted species (42).

Method

1. Dilute 20–25 µL of serum with an equal volume of 10 mM Tris–HCl, 150 mM sodium chloride, pH 7.4
2. Centrifuge the diluted sample at $2,500 \times g$ to remove possible small particles in suspension.

3. Inject the sample into the immunosorbent column according to the instructions provided by the supplier.
4. Collect the column effluent at the outlet of the column.
5. Wash the column with approximately 3–5 volumes of washing buffer (per supplier instructions).
6. Repeat steps 1–5 if necessary to generate sufficient sample. Often several depletion cycles are required to generate sufficient yield for downstream applications.
7. Regenerate the column by injecting about three volumes of stripping buffer followed by about 10 volumes of neutralization buffer.

The collected protein sample is always diluted by both the passage through the column and the washing step. Depending on supplier recommendations, the dilution factor can range from 5- to more than 150-fold. With most diluted treated samples, a concentration step is required prior to downstream analysis. Available processes to concentrate the sample include dialysis followed by lyophilization, ultrafiltration, and protein precipitation. Although all of them are routinely used, they all suffer from several drawbacks such as losses during dialysis, nonspecific binding on filtration membranes, and incomplete precipitation of low mass proteins. In addition to concentration, other adjustments may be required depending on the downstream analysis method to be used (see Note 1). Depleted samples are often subjected to additional purification steps to further reduce sample complexity (see Note 2).

Table 1 reports experimental data on several depletion methods.

3.1.2. Reduction of Dynamic Protein Concentration Range

As opposed to depletion, this process consists of "compressing" the dynamic range of protein concentrations by simultaneously diluting the high-abundance species and concentrating trace proteins. The reduction of dynamic protein concentration range is based on the use of a solid-phase mixture of an extremely large number of specific sorbents, each of them carrying a given hexapeptide ligand (5). Proteins contacted with such a mixture of affinity media are captured by their respective partner ligand under physiological conditions, until saturation is reached. Proteins present at a high concentration rapidly saturate their partner bead ligands so that the vast majority of the same protein remains unbound. Conversely, low-abundance proteins concentrate on their ligand up to saturation as long as the load is sufficient. On the basis of the saturation-overloading chromatographic principle, a solid-phase ligand library enriches for trace proteins while concomitantly reducing the concentration of abundant species. Desorption of captured species is operated by classical eluting agents used in affinity chromatography as summarized in Table 2; sodium dodecyl

Table 1
Main characteristics of high-abundance protein depletion methods

Removed proteins	Affinity ligands	Loading conditions	Dilution factor[a]
Human serum albumin	Cibacron Blue	20 mM phosphate buffer pH 7.0	20
Human serum albumin and IgG	Cibacron Blue and protein A	10 mM Tris–HCl, pH 7.5 + 25 mM NaCl	10
Serum immunoglobulins G	Protein G	Phosphate buffered saline	5
Human serum albumin	Antibodies	Phosphate buffered saline	180
Human serum albumin and IgG	Antibodies	Phosphate buffered saline	35
Humans serum albumin and IgG	Specific ligands	Phosphate buffered saline	30
Six abundant human serum proteins	Antibodies	10 mM Phosphate pH 7 + 0.5 M NaCl	60

[a]Dilution obtained using the recommended protocol from the vendor

Table 2
Protein desorbing solutions for the use with peptide libraries

Possible elution solution	Applicability	Mechanisms, effects and compatibility
5–10% sodium dodecyl sulfate	Single elution, denaturing	Desorbs all proteins. The method is compatible with SDS-PAGE, but not with isoelectric focusing migrations and mass spectrometry
6 M guanidine-HCl, pH 6	Single elution, denaturing	Dissociates all types of interactions; proteins are fully denatured. Proteins need to be precipitated for further analytical methods, except for RPC separation
9 M urea, 2% CHAPS, citric acid to pH 3.0–3.5	Single elution, partial denaturation	Dissociates most interactions except very strong hydrophobic associations. Fully compatible with isoelectric focusing, 2DE and SELDI (if neutralized and fivefold diluted). Otherwise proteins need to be precipitated for further analytical methods
9 M urea, 2.4% CHAPS, ammonia to pH 11	Single elution, partial denaturation	Dissociates most interactions except very strong hydrophobic associations. Fully compatible with isoelectric focusing, 2DE and SELDI (if neutralized and fivefold diluted). Otherwise proteins need to be precipitated for further analytical methods. The eluate requires neutralization

(continued)

Table 2 (continued)

Possible elution solution	Applicability	Mechanisms, effects and compatibility
0.5–1 M sodium chloride	Fractionated elution	Desorbs proteins that are captured by a dominant ion exchange effect. Compatible with certain types of liquid chromatography such as HIC, RPC, IMAC
200 mM glycine–HCl, pH 2.4	Fractionated elution	Desorbing solution for the dissociation of biological affinity interactions (e.g., Ag–Ab). Proteins need to be precipitated for further analytical methods, except for SELDI and RPC analysis. The eluate requires neutralization
50% ethylene glycol in water	Fractionated elution	Dissociates mild hydrophobic interactions. Compatible with most analytical methods except HIC and RPC
Acetonitrile (6.6)–isopropanol (13.3)–trifluoroacetic acid (0.8)–water(79.3)	Fractionated elution	Dissociates strong hydrophobic interactions. Compatible with most analytical methods except HIC and RPC, unless solvents are evaporated. The eluate requires neutralization
7 M urea–2 M thiourea–4% CHAPS	Fractionated elution	Dissociates mixed modes, hydrophobic associations and hydrogen bonding. Fully compatible with isoelectric focusing, 2DE and SELDI.

sulfate solution for electrophoresis (e.g., Laemli buffer) could be used for an exhaustive protein desorption (43).

An example of human serum treated with ProteoMiner is illustrated on Fig. 1 where elution was performed in two sequential steps. Analyses of eluates were performed using two-dimensional electrophoresis and SELDI MS.

Since this treatment is based on thermodynamic principles, its performance depends on the combinations of dissociation constants and the concentration of very dilute proteins. Although the large diversity of peptide ligands from the library covers all necessary affinity interactions to capture the proteins present in the sample, its reproducibility is ensured by a rigorous control of environmental pH, ionic strength and temperature. In the case of comparative profiling, all different operations must be maintained rigorously identical and the ratio between the bead volume and the sample volume are to be calculated so that the beads are overloaded (44).

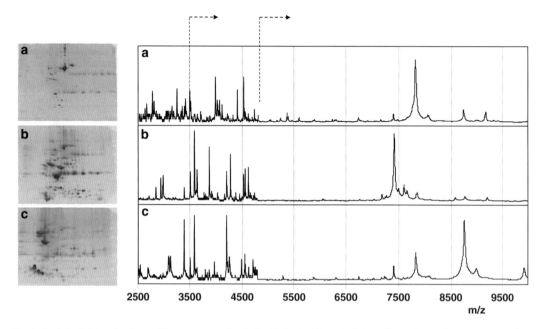

Fig. 1. Analytical determinations of human serum treated with ProteoMiner. *Left*: two-dimensional electrophoresis; *right*: SELDI MS. (**a**) Initial untreated human serum; (**b**) bead library eluate using 7 M urea containing 2 M thiourea and 2% CHAPS; (**c**) eluate using 9 M urea containing 2% CHAPS and 50 mM citric acid. Two-dimensional electrophoresis was performed using a pH gradient between 3 and 10. SELDI MS analysis was performed using an IMAC chip; the mass range shown is from 2.5 to 10 kDa.

Method

1. Wash repeatedly 100 μL of a hexapeptide ligand library (e.g., ProteoMiner) with PBS either in a spin column or in a small chromatographic packed column.

2. Load the clear protein extract (e.g., 1 mL human serum) and incubate for 60 min at room temperature under gentle shaking or using a downward column flow. In case of other protein extracts, the amount of proteins to be injected should be at least 30–50 mg total. In case of proteases are suspected to be present, add appropriate inhibitors such as protease inhibitor cocktail.

3. Wash out the excess of proteins from the hexapeptide ligand library by either centrifugation or column washing depending on the mode of protein contact.

    (a) For spin columns, the excess of proteins is removed by centrifugation at about $1,250 \times g$ followed by at least three washings with ten bead volumes of the initial physiological buffer. In between centrifugations the bead slurry is shaken gently for 10 min.

    (b) When using a chromatographic column, pump through 10 column volumes of PBS to ensure that all protein excess is eliminated (baseline signal at 280 nm).

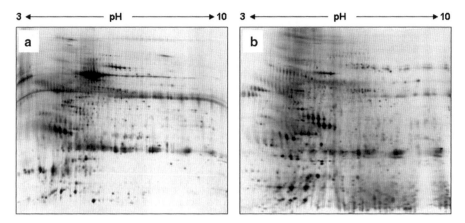

Fig. 2. Two-dimensional electrophoresis of urinary proteome before (**a**) and after (**b**) a treatment with ProteoMiner to reveal low-abundance proteins. Protein desorption was performed using 10% sodium dodecyl sulfate. Courtesy of Dr G. Candiano, Gaslini Institute, Genova (44).

4. Elute proteins captured by the hexapeptide ligand library beads, either by one single operation or by a sequence of elution steps. A single elution is performed using the urea-CHAPS-acetic or citric acid solution. Desorb captured proteins using three successive steps with 1 volume of elution solution.

   a. Between elution steps, shake the suspension for 10 min at room temperature.

   b. Centrifuge at $1,250 \times g$.

The collected pooled eluates contain captured proteins from the initial sample and can be directly analyzed by SELDI-TOF MS following dilution into the appropriate binding buffer. Multi-step elution schemes, which fractionate bound proteins using three to four sequential elution buffers, have also been used routinely for SELDI analysis. This method is performed as described above except that a washing sequence with different eluting agents is applied sequentially. A variety of elution conditions can be used depending on the purpose and are summarized in Table 2 (see Notes 3 and 4). When collected fractions are acidic or alkaline they must be immediately neutralized. Eluted fractions are used for proteomics studies. Figure 2 shows a comparative example of the urinary proteome before and after treatment with a combinatorial hexapeptide ligand library.

3.2. Fractionation Approaches with Solid-Phase Media

Chromatographic separation of proteins from complex mixtures is the result of interactions with solid phases followed by sequential specific or nonspecific desorption. A large number of protein–solid phase interactions are known in the field of proteomics investigations (29). They are based on physicochemical properties of proteins to separate or on biochemical recognition. In addition to the different mechanisms of action, various modes of exploitations are also possible such as packed beds, suspension separation, or even fluidized bed fractionations. Multiple columns with complementary properties are used to enhance the capability of chromatography to separate and isolate gene products. The field of liquid chromatography of proteins is extremely large and it is beyond the scope of this chapter to review all practical possibilities. Instead, we will describe protein separations methods that have been used successfully in the SELDI-TOF proteomics workflow. The main driving force of these methods is to find modes of separations that are the most powerful to decrease protein redundancy between a reduced number of collected fractions. Reduced redundancy of proteins in collected fractions is a way to reduce the number of fractions necessary to maintain a high level of analytical resolution and hence the power of discovery. Reducing the number of fractions required is of utmost importance, especially when a very large number of samples has to be analyzed comparatively with a reasonable workload. Other objectives include selecting conditions that yield the best complementary behavior between solid phases, the minimum dilution and the absence of polypeptide losses due to, for instance, nonspecific binding on solid phases.

3.2.1. Classical Chromatography Separation Using Fractionated Elution

Chromatography is the most popular way to fractionate protein mixtures with the objective to simplify the analysis of components in proteomics investigations. A number of reviews have focused on liquid chromatography for protein fractionation exploiting various physicochemical interaction principles (29, 45–48). The process of liquid chromatography could be described as an adsorption of proteins on a solid-phase media followed by a fractionated elution of proteins using progressively more stringent desorbing agents. Many fractionation strategies use one or more of three generic separation methods: (1) ion exchange chromatography, which separates proteins based on protein charges, (2) gel filtration, which separates based on differences in protein masses and (3) hydrophobic associations between lipophilic sites of proteins with hydrocarbon chemical functions attached on the surface of chromatography sorbents. Outside these generic fractionation possibilities, there are more selective adsorptions addressing groups of proteins or even single proteins.

Within the proteomics world, Fountoulakis' group has already in 1998 extensively developed and described chromatography approaches as ways to simplify complex proteomes (49). In a typical example, affinity chromatography on a heparin affinity sorbent was used as a prefractionation step for enriching certain protein fractions.

Hydrophobic, cationic and anionic proteins are easily fractionated using classical liquid chromatography methods. On the contrary, affinity chromatography (50) can be used to enrich for classes of proteins based on specific interaction properties, such as glycoproteins using lectins of different specificities (30, 33, 51), or phosphoproteins via metal chelating resins (21), affinity for zirconium or titanium oxide (52, 53) or proteins interacting with heparin (54) just to mention a few.

While easy to implement and quite well known, one of major drawbacks of liquid chromatography, especially when performed in a 96-well filtration plate format with a large number of samples and low number of fractions, is protein tailing. This results in the presence of individual proteins in several fractions along with a significant dilution of proteins within each fraction. Another limitation is the dilution effect of collected proteins.

As many different combinations of samples and chromatography types are currently used, this paragraph limits its scope to the most popular protein fractionation of human serum by anion exchange chromatography for proteomics analysis (19).

Method

1. Fill a spin column (or a well of a 96-well filtration plate) with 125 mL of anion exchange solid-phase media (e.g., Q-HyperD).
2. Equilibrate column in binding buffer (composed of 1 M urea-0.22% CHAPS, 50 mM Tris–HCl, pH 9).
3. Separately, mix 70 µL of human serum with an equal volume of a solution composed of 9 M urea and 2% CHAPS and gently stir for 60 min at room temperature.
4. Once the column is equilibrated, add the diluted serum to the solid phase and incubate for 30 min.
5. Collect the flow-through fraction including a wash with 140 µL of U9 (250 µL).
6. Elute captured proteins in a stepwise manner with buffers of decreasing pH:
    (a) 50 mM HEPES, pH 7. Use 140 µL eluting solution each step.
    (b) 100 mM sodium acetate pH 5.
    (c) 100 mM sodium acetate pH 4.
    (d) 50 mM sodium citrate pH 3.
    (e) Finally, the solid phase is stripped using a mixture constituted of 33.3% isopropanol, 16.6% acetonitrile, 0.1% trifluoroacetic acid in DI water).

    In practice, in each step the solid-phase media added with the elution solution, is incubated for 30 min and then the supernatant is collected by centrifugation or filtration.
7. The six protein fractions obtained are compatible with further proteomics analysis using SDS-PAGE, two-dimensional electrophoresis or SELDI-MS.

24 L. Guerrier et al.

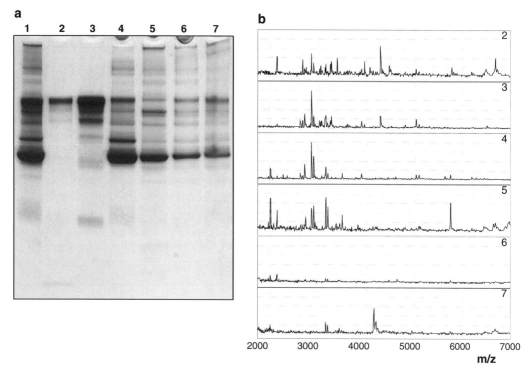

Fig. 3. Analysis of human serum polypeptide species separated by anion exchange chromatography by SDS-PAGE (high masses "**a**") and by SELDI MS (low masses "**b**"). (**a**) SDS-PAGE analysis of human serum fractions obtained from Q-HyperD fractionation in view of protein profiling with SELDI MS. The fractionation was performed using a stepwise decrease of pH and after having treated the initial serum proteins with urea. *Lane 1*: initial human serum; *Lane 2*: flowthrough fraction pH 9.0; *Lane 3*: elution at pH 7.0; *Lane 4*: elution at pH 5.0; *Lane 5*: elution at pH 4.0; *Lane 6*: elution at pH 3.0; *Lane 7*: elution with a hydro-organic mixture. (For details see Subheading 3.1.2). (**b**) SELDI MS associated with CM10 ProteinChip arrays of fractions obtained as indicated above. *m/z*= mass over charge.

Figure 3 shows a typical analytical result after SDS-PAGE and SELDI MS on CM10 ProteinChip array.

3.2.2. Fractionation with Stacked Sorbents and Single Buffer

A novel method for protein fractionation from complex mixtures is the use of superimposed chromatographic sorbents as a cascade with a single loading buffer and a single elution solution to recover captured proteins (55). This system is composed of a set of solid-phase media serially connected in a stacked mode, equilibrated in the same binding buffer. The protein sample is applied at the top and it crosses all the solid-phase layers, each of them capable of capturing a subset of proteins, so that, when the sample reaches the bottom-positioned media, most if not all proteins are removed. Once the protein adsorption phase is completed, the sequence of stacked beds is washed with the equilibration buffer until it reaches baseline (as measured at 280 nm). Then each stacked section is dissociated and proteins desorbed independently from each sectional column. In essence, this resembles a series of depletion columns, each of them in charge of removing a given group of proteins.

Based on the selection of orthogonal solid-phase chemistries and their sequence, bound proteins are different in each of the capturing blocks. Naturally the process should be used in such a way so that sectional columns are not overloaded with proteins that are supposed to be captured, so as to minimize or eliminate the level of protein redundancy (the presence of a same protein in different fractions) between fractions. When superimposing solid-phase layers, media are assembled so that the top comprises the most selective material, and the least selective media is positioned at the bottom of the sequence.

It has been reported that for a same limited number of fractions (e.g., five or six), the total number of species observed by mass spectrometry was more than doubled compared to a classical elution chromatography (56).

Such an approach, when coupled with high throughput capability, appears to be useful for discovering protein species with diagnostic relevance. Only a limited number of fractions is generated and the separation process can be operated very rapidly even with a large number of samples. Interestingly, species that are relatively diluted are captured only by one column section and are consequently concentrated and therefore better detected in comparison to classical methods.

The selection of solid-phase media can be randomized or can be adapted to well-defined fractionations. For instance, for glycoprotein fractionation, lectin media could be used (concanavalin A, the least selective would be positioned in the last position). Chelating media loaded with different metal ions could also be used or different hydrophobic solid-phase media where the least hydrophobic will be positioned first.

Figure 4 shows the separation of species captured by a cascade of chelated metal ions immobilized on IMAC resins, as analyzed by SELDI MS.

Although this approach is quite different from the use of Immunosorbents for "depletion" of high-abundance species (see Subheading 2.1.1), a cascade of immunosorbents could also be envisioned.

Detailed methods for fractionation of serum using stacked sorbents are described below. These methods can be modified for other sample types and binding specificities.

Method

1. In order to prevent protein–protein interactions and formation of aggregates, partially denature serum from the sample prior to loading on the resin. Add an equal volume of a solution composed of 9 M urea and 2% CHAPS to the human serum sample and incubate for 60 min at room temperature.

2. Adjust the denatured serum solution, adding 36 µL of a 1 M Tris–HCl buffer pH 9, 100 µL of a 9 M urea-2% CHAPS solution and 364 µL of deionized water for every 0.4 mL of initial protein solution.

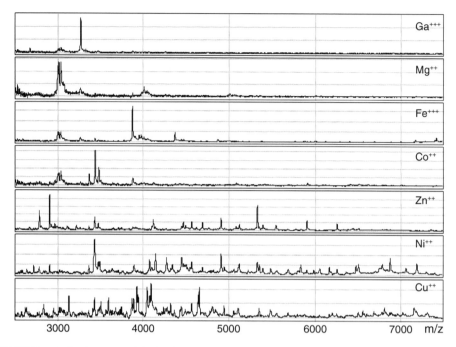

Fig. 4. SELDI MS analysis of serum protein fractions obtained from a cascade of IMAC columns loaded with metal ions as indicated in the *boxes* (from Ga^{3+} to Cu^{2+}). Loading was performed in physiological buffer and elution was obtained by decreasing the pH to 4.0. Each fraction was analyzed using a CM10 ProteinChip array. The visualized mass window is between 2,500 and 7,500 Da.

3. Superimpose columns in the following order from the top to the bottom: Immobilized Protein A, Affigel Cibacron Blue, Heparin-agarose, MEP-HyperCel, Green-5 agarose and PPA-HyperCel. Volumes of each solid-phase media should be 150 µL. Volumes of Protein A and Cibacron Blue resins may need to be adjusted depending on the sample type and resin capacity (see Note 5).

4. Equilibrate superimposed media columns with a binding buffer composed of 16 volumes of phosphate buffered saline, 9 volumes of 1 M Tris–HCl buffer, pH 8 and 75 volumes of distilled water. The operation is performed using a peristaltic pump at a flow-rate of 0.2 mL/min.

5. Following the equilibration step, inject 160 µL of the denatured serum sample to be fractionated at the top of the column at a flow rate of 0.01 mL/min followed by the binding buffer at the same flow rate to push the sample throughout the series of sorbents.

6. The first 1,250 µL of effluent is discarded and the next 1,250 µL effluent is collected as the flow-though fraction.

7. Separate columns and elute adsorbed proteins using 500 µL of desorbing solution (see Note 6).

Many variants of this general method are possible especially when using other solid-phase media. An example is given in Fig. 3 where the purpose was to separate proteins having affinity for different metal ions.

3.3. pI Separation Groups by Solid-State Buffers

A specialized variant of fractionation with stacked sorbents is the separation of proteins according to their isoelectric point using so-called "Solid-State Buffers" (SSB) associated with ion exchangers (35). This approach contrasts with current methods for separation by isoelectric point since it uses chromatographic-like processes compared to electrically driven separation. Advantages of this approach include the ability to perform separation rapidly without ampholine carriers and to use sample volumes that can be of undefined volume and concentration.

Stacked columns are mixed beds of cation or anion exchange media mixed in equal volumes with SSB capable of spontaneously generating predetermined pHs (35, 56). Solid-State Buffers are discrete beads of amphoteric polymers inducing a predetermined environmental pH in the presence of ionizable molecules such as proteins. SSB-IEX mixtures are packed in distinct columns under a cascade configuration as described in the previous section. When the packed media blend comprises an anion exchange resin, the top sectional column contains the SSB of the lowest pH value and the bottom sectional column contains the SSB with the highest pH value. On the contrary, when a cation exchanger is mixed with SSBs, an inverse pH sequence is to be configured. A sample of the protein extract solution to fractionate is introduced at the top of the first column. Proteins and small ions present in the solution enter in contact with the SSB, which generates a well-defined environmental pH, influencing thus the net charge of proteins present. Those having a net charge complementary to the ion exchange resin are captured and the others are progressively pushed in the second sectional column. In this second mixed media environment, another SSB induces a different environmental pH generating consequently a different net charge of upcoming proteins from the first sectional column. Here also proteins of complementary charges are attracted by the ion exchanger while proteins with antagonistic charges are repelled and pushed into the following sectional column and so on. The process is then repeated a number of times equivalent to the number of sectional columns. Once the injection of sample and washings are complete, sectional columns are dissociated and adsorbed proteins from each mixed bed are eluted and collected separately by increasing ionic strength or changing pH or both. The collected fractions comprise proteins of pI ranges depending on the pH induced by the local SSB used.

To avoid the effects of ion interferences and prevent protein–protein interactions, the protein sample is solubilized in low ionic strength solutions containing dissociating agents such as urea or is diluted two to five times with a similar solution. It has been recently demonstrated that human serum proteome is better fractionated using Solid-State Buffers compared to classical anion exchange chromatography (57).

3.3.1. Method

1. Prepare SSB-IEX blends by mixing equal volumes of SSB with either a Q anion exchanger or a CM ion exchanger. For optimal performance acidic proteins should be separated using media blends comprised of cation exchangers and alkaline proteins with anion exchanger blends.
   (a) The pHs induced of selected SSBs are 4.6, 5.4, 6.2, 7.0, 8.5, and 9.3.
   (b) Wash each mixture with a liquid buffer of a pH identical to the SSB and then exhaustively rinse with water or 2 mM potassium chloride containing 4 M urea, to eliminate all components of the buffers.
2. Pack the conditioned bead mixtures into distinct small columns (e.g., 6 mm diameter and 3 cm long) and connect to each other according in an appropriate order.
   (a) When using an anion exchanger, connect columns in order of increasing pH.
   (b) When using a cation exchanger, connect columns in order of decreasing pH.
3. Inject the protein sample (previously dissolved in 4 M urea containing 2–5 mM potassium chloride) into the first column and wash with 5–10 column volumes of urea-KCl solution to remove all noncaptured proteins (see Note 7).
4. Disconnect columns and elute bound proteins separately using a solution of 1 M sodium chloride or a change of pH to ensure proper dissociation of proteins from the ion exchanger.
5. Collected proteins are ready for proteomics studies (see Note 8). In case of incompatibility of pH and/or ionic strength, proteins can be desalted according to currently available methods.

Figure 5 illustrates an example of serum protein separation within the acidic region.

4. Notes

1. Protein samples obtained using depletion processes are equilibrated with the initial buffer (see details on Table 1). Samples are generally at neutral pH and the buffer is often PBS. Under these conditions the sample is neither directly compatible with

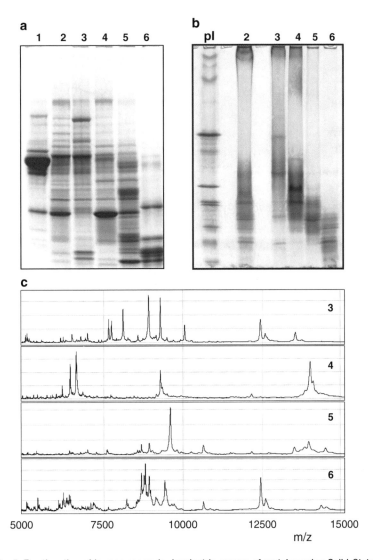

Fig. 5. Fractionation of human serum by isoelectric groups of proteins using Solid-State Buffers. Human serum was first treated with ProteoMiner to enhance the presence of low-abundance proteins and reduce the concentration of high-abundance ones. The fractionation was performed using SSB of pH 4.6, 5.4 and 6.2 associated with a CM cation exchange media. (a) SDS-PAGE analysis of produced fractions. (b) IEF analysis of same fractions. (c) SELDI MS associated with CM10 ProteinChip array for the mass analysis of separated fractions. *Lane 1*: Initial human serum; *Lane 2*: Serum treated with ProteoMiner; *Lane 3*: Flowthrough fraction or proteins of pI > to 6.2; *Lane 4*: Protein fraction of pI between 5.4 and 6.2; *Lane 5*: Protein fraction of pI between 4.6 and 5.4; *Lane 6*: Protein fraction of pI below 4.6. pI: markers of isoelectric point. Proteins from each fraction are complementary each other.

analytical two-dimensional electrophoresis nor with direct MALDI mass spectrometry. It can, however, be used for LC-MS/MS or SELDI-MS analysis after simple dilution and chip washing.

2. Several options are available if further fractionation is desired to simplify the protein content. Samples can be used for ion exchange fractionation after diluting at least tenfold and adjusting pH to 4.5–6.0 for cation exchange chromatography or 7.5–9.0 for anion exchange chromatography. Fractionation by isoelectric focusing requires a complete desalting.

3. Although there are several methods to desorb proteins en masse, an interesting way to quantitatively collect all captured proteins is to use 5–10% SDS in water. However, the presence of SDS may or may not be compatible with analytical methods to follow, so that in some cases the elimination of SDS is necessary (for more details see ref. 43). When the following analysis is a determination of a biological activity or an interaction with antibodies such as immunoblots or ELISA-based assays, non-denaturing solutions should be tried.

4. As opposed to one single elution, there is the possibility to desorb proteins sequentially to further separate proteins based on their physicochemical characteristics and thus simplify the following analysis (5).

5. The most critical point is to avoid overloading each individual solid-phase media. Therefore an excess if resin is recommended even when the binding capacity is known. In the case of serum, where albumin and IgG dominate, it is advisable to use a relatively large volume of Cibacron resin (about three times compared to other resins) and to select Protein A media of high binding capacity.

6. These chromatographic media are theoretically reusable after proper cleaning performed according to the recommendations of vendors; however, it is advised not to reuse the resins or to limit the number of reutilizations to prevent binding capacity losses and thus a possible overloading situation as well as cycle to cycle carryover.

7. Although less critical than in isoelectric focusing, the initial ionic strength of the protein extract to fractionate may be important for a proper fractionation without saturation of ion exchange resin. If the ionic strength of the initial protein solution is considered high, it is advised to dilute with water or better with a solution of 4 M urea even if this results into a large volume of sample to treat. The method in fact is perfectly adapted to large volumes of dilute proteins without interfering with the final fractionation results.

8. Obtained pI cuts may not be extremely precise at the expected pH value. This is not critical because the primary goal is to obtain fractions that contain complementary proteins with minimal redundancy.

References

1. Anderson, L.N. and Anderson, N.G. (2002) The human plasma proteome. History, character and diagnostic prospects. *Mol. Cell. Proteomics.* **1**, 845–867.
2. Greenough, C., Jenkins, R.E., Kitteringham, N.R., Pirmohamed, M., Park, B.K., Pennington, S.R. (2004) A method for the rapid depletion of albumin and immunoglobulin from human plasma. *Proteomics.* **4**, 3107–3111.
3. Dardé, V.M., Barderas, M.G., Vivanco, F. (2007) Depletion of high-abundance proteins in plasma by immunoaffinity subtraction for two-dimensional difference gel electrophoresis analysis. *Methods Mol Biol.* **357**, 351–364.
4. Liu, T., Qian, W.J., Mottaz, H.M., Gritsenko, M.A., Norbeck, A.D., Moore, R.J., Purvine, S.O., Camp, D.G., Smith, R.D. (2006) Evaluation of multiprotein immunoaffinity subtraction for plasma proteomics and candidate biomarker discovery using mass spectrometry. *Mol. Cell. Proteomics.* **5**, 2167–2174.
5. Thulasiraman, V., Lin, S., Gheorghiu, L., Lathrop, J., Lomas, L., Hammond, D., Boschetti, E. (2005) Reduction of the concentration difference of proteins in biological liquids using a library of combinatorial ligands. *Electrophoresis.* **26**, 3561–3571.
6. Sennels, L., Salek, M., Lomas, L., Boschetti, E., Righetti, P.G., Rappsilber, J. (2007) Proteomic Analysis of Human Blood Serum Using Peptide Library Beads. *J. Proteome Res.* **6**, 4055–4062.
7. Guerrier, L., Claverol, S., Fortis, F., Rinalducci, S., Timperio, A.M., Antonioli, P., Jandrot-Perrus, M., Boschetti, E., Righetti, P.G. (2007) Exploring the Platelet Proteome via Combinatorial, Hexapeptide Ligand Libraries. *J. Proteome Res.* **6**, 4290–4303.
8. Roux-Dalvai, F., Gonzalez de Peredo, A., Simó, C., Guerrier, L., Bouyssie, D., Zanella, A., Citterio, A., Burlet-Schiltz, O., Boschetti, E., Righetti, P.G, Monsarrat, B. (2008) Extensive analysis of the cytoplasmic proteome of human erythrocytes using the peptide ligand library technology and advanced mass spectrometry. *Mol. Cell Proteomics.* **7**, 2254–2269.
9. Herbert, B.R., Righetti, P.G., Citterio, A., Boschetti, E. (2007) Sample preparation and pre-fractionation techniques for electrophoresis-based proteomics, in *Proteome Research: Concepts, Technology and Applications* (Wilkins, M.R., Appel, R.D., Williams, K.L., Hochstrasser, D.F., eds.) Springer, Berlin, pp. 15–40.
10. Huber, L.A., Pfaller, K., Vietor, I. (2003) Organelle proteomics: implications for subcellular fractionation in proteomics. *Circulation Res.* **92**, 962–968.
11. de Araùjo, M.E., Huber, L.A., Stasyk, T. (2008) Isolation of endocytic organelles by density gradient centrifugation. *Methods Mol Biol.* **424**, 317–331.
12. Gauthier DJ, Lazure C. (2008) Complementary methods to assist subcellular fractionation in organellar proteomics. *Expert Rev Proteomics.* **5**, 603–617
13. Bae, S.H., Harris, A.G., Hains, P.G., Chen, H., Garfin, D.E., Hazell, S.L., Paik, Y.K., Walsh, B.J., Cordwell, S.J. (2003) Strategies for the enrichment and identification of basic proteins in proteome projects. *Proteomics.* **3**, 569–579.
14. Mertens-Strijthagen, J., De Schrijver, C., Wattiaux-De Coninck, S., Wattiaux, R. (1979) A centrifugation study of rat-liver mitochondria, lysosomes and peroxisomes during the perinatal period. *Eur J Biochem.* **98**, 339–352.
15. Reschiglian, P., Moon, M.H. (2008) Flow field-flow fractionation: a pre-analytical method for proteomics. *J. Proteomics.* **71**, 265–276.
16. Herbert, B., Righetti, P.G. (2000) A turning point in proteome analysis: sample prefractionation via multicompartment electrolyzers with isoelectric membranes. *Electrophoresis.* **21**, 3639–3648.
17. Tang, H.Y., Speicher, D.W. (2005) Complex proteome prefractionation using microscale solution isoelectrofocusing. *Expert Rev Proteomics.* **2**, 295–306.
18. Michel, P.E., Crettaz, D., Morier, P., Heller, M., Gallot, D., Tissot, J.D., Reymond, F., Rossier, J.S. (2005) Proteome analysis of human plasma and amniotic fluid by Off-Gel isoelectric focusing followed by nano-LC-MS/MS. *Electrophoresis.* **27**, 1169–1181.
19. Wiesner, A. (2004) Detection of tumor markers with ProteinChip technology. *Current Pharm. Biotechnol.* **5**, 45–67.
20. Tirumalai, R.S., Chan, K.C., Prieto, D.A., Issaq, J., Conrads, T., Veenstra, T.D. (2003) Characterization of the low molecular weight human serum proteome. *Mol. Cell Proteomics.* **2**, 1096–1103.
21. Fountoulakis, M., Takács, M.F., Takács, B. (1999) Enrichment of low-copy-number gene products by hydrophobic interaction chromatography. *J Chromatogr A.* **833**, 157–168.
22. Shen, H., Li, X., Bieberich, C.J., Frey, D.D. (2008) Reducing sample complexity in proteomics by chromatofocusing with simple buffer mixtures. *Methods Mol Biol.* **424**, 187–203.

23. Opiteck, G.J., Ramirez, S.M., Jorgenson, J.W., Moseley, M.A. 3rd. (1998) Comprehensive two-dimensional high-performance liquid chromatography for the isolation of overexpressed proteins and proteome mapping. *Anal Biochem.* **258**, 349–361.
24. Sun, X., Chiu, J.F., He, Q.Y. (2005) Application of immobilized metal affinity chromatography in proteomics. *Expert Rev Proteomics.* **2**, 649–657.
25. Righetti, P.G., Castagna, A., Antonioli, P., Boschetti, E. (2005) Prefractionation techniques in proteome analysis: the mining tools of the third millennium. *Electrophoresis.* **26**, 297–319.
26. Cañas, B., Piñeiro, C., Calvo, E., López-Ferrer, D., Gallardo, J.M. (2007) Trends in sample preparation for classical and second generation proteomics. *J Chromatogr A.* **1153**, 235–258.
27. Tam, S.W., Pirro, J., Hinerfeld, D. (2004) Depletion and fractionation technologies in plasma proteomic analysis. *Expert Rev Proteomics.* **1**, 411–420.
28. Matt, P., Fu, Z., Fu, Q., Van Eyk, J.E. (2008) Biomarker discovery: proteome fractionation and separation in biological samples. *Physiol Genomics.* **33**, 12–17.
29. Jmeian, Y., El-Rassi, Z. (2009) Liquid-phase-based separation systems for depletion, prefractionation and enrichment of proteins in biological fulids for in depth proteomics analysis. *Electrophoresis.* **30**, 249–261.
30. Qiu, R., Regnier, F.E. (2005) Use of multidimensional lectin affinity chromatography in differential glycoproteomics. *Anal. Chem.* **77**, 2802–2809.
31. Plavina, T., Wakshull, E., Hancock, W.S., Hincapie, M. (2007) Combination of abundant protein depletion and multi-lectin affinity chromatography (M-LAC) for plasma protein biomarker discovery. *J Proteome Res.* **6**, 662–671.
32. Hirabayashi, J. (2004) Lectin-based structural glycomics: glycoproteomics and glycan profiling. *Glyconj. J.* **21**, 35–40.
33. Dayarathna, M.K., Hancock, W.S., Hincapie, M. (2008) A two step fractionation approach for plasma proteomics using immunodepletion of abundant proteins and multi-lectin affinity chromatography: Application to the analysis of obesity, diabetes, and hypertension diseases. *J. Sep. Sci.* **31**, 1156–1166.
34. Li, Y., Leng, T., Lin, H., Deng, C., Xu, X., Yao, N., Yang, P., Zhang, X., (2007) Preparation of Fe_3O_4-ZrO_2 core-shell microspheres as affinity probes for selective enrichment and direct determination of phosphopeptides using matrix-assisted laser desorption ionization mass spectrometry. *J. Proteome Res.* **6**, 4498–4510.
35. Fortis, F., Guerrier, L., Girot, P., Fasoli, E., Righetti, P.G. Boschetti, E. (2008) A pI-based protein fractionation method using Solid-State Buffers. *J. Proteomics.* **71**, 379–389.
36. Guerrier, L., Boschetti, E. (2007) Protocol for the purification of proteins from complex biological extracts for identification by mass spectrometry. *Nature Protocols.* **2**, 831–837.
37. Bhattacharya, D., Mukhopadhyay, D., Chakrabarti, A. (2007) Haemoglobin depletion from red blood cell cytosol reveals new proteins in 2-D gel-based proteomics study. *Proteomics Clin. Appl.* **1**, 561–564.
38. Gao, M., Deng, C., Yu, W., Zhang, Y., Yang, P., Zhang, X. (2008) Large scale depletion of the high-abundance proteins and analysis of middle- and low-abundance proteins in human liver proteome by multidimensional liquid chromatography. *Proteomics.* **8**, 939–947.
39. Cellar, N.A., Kuppannan, K., Langhorst, M.L., Ni, W., Xu, P., Young, S.A. (2008) Cross species applicability of abundant protein depletion columns for ribulose-1,5-bisphosphate carboxylase/oxygenase. *J Chromatogr B Analyt Technol Biomed Life Sci.* **861**, 29–39.
40. Harezlak, J., Wang, M., Christiani, D., Lin, X., (2007) Quantitative quality-assessment techniques to compare fractionation and depletion methods in SELDI-TOF mass spectrometry experiments. *Bioinformatics.* **23**, 2441–2448.
41. Echan, L.A., Tang, H.Y., Ali-Khan, N., Lee, K., Speicher, D.W. (2005) Depletion of multiple high-abundance proteins improves protein profiling capacities of human serum and plasma. *Proteomics.* **5**, 3292–3303.
42. Righetti, P.G., Boschetti, E., Lomas, L., Citterio, A. (2006) Protein Equalizer technology: The quest for a "democratic proteome". *Proteomics.* **6**, 3980–3992.
43. Candiano, G., Dimuccio, V., Bruschi, M., Santucci, L., Gusmano, R., Boschetti, E., Righetti, P.G., Ghiggeri, G.-M. 2009Combinatorial peptide ligand libraries for urine proteome analysis: investigation of different elution systems. *Electrophoresis.* **30**, 2405–2411.
44. Fasoli, E., Farinazzo, A., Sun, C.J., Kravchuck, A.V., Guerrier, G., Fortis, F., Boschetti, E., Righetti, P.G. (2010) Interaction among proteins and peptide libraries in proteome analysis: pH involvement for a larger capture of species. *J. Proteomics.* **73**, 733–742.
45. Stasyk, T., Huber, L.A. (2004) Zooming in: fractionation strategies in proteomics. *Proteomics.* **4**, 3704–3716.

46. Lee, H.J., Lee, E.Y., Kwon, M.S., Paik, Y.K. (2006) Biomarker discovery from the plasma proteome using multidimensional fractionation proteomics. *Curr Opin Chem Biol.* **10**, 42–49.
47. Nice, E.C., Rothacker, J., Weinstock, J., Lim, L., Catimel, B. (2007) Use of multidimensional separation protocols for the purification of trace components in complex biological samples for proteomics analysis. *J Chromatogr A.* **1168**, 190–210.
48. Matt, P., Fun Z., Fun Q., Van Eykn J.E. (2008) Biomarker discovery: proteome fractionation and separation in biological samples. *Physiol Genomics.* **33**, 12–17.
49. Fountoulakis, M., Takács, B. (1998) Design of protein purification pathways: application to the proteome of Haemophilus influenzae using heparin chromatography. *Protein Expr Purif.* **14**, 113–119
50. Azarkan, M., Huet, J., Baeyens-Volant, D., Looze, Y., Vandenbussche, G. (2007) Affinity chromatography: a useful tool in proteomics studies. *J Chromatogr B.* **849**, 81–90.
51. Geyer, H., Geyer, R. (2006) Strategies for analysis of glycoprotein glycosylation. *Biochim Biophys Acta.* **1764**, 1853–1869.
52. Feng, S., Ye, M., Zhou, H., Jiang, X., Zou, H., Gong, B. (2007) Immobilized zirconium ion affinity chromatography for specific enrichment of phosphopeptides in phosphoproteome analysis. *Mol. Cell. Proteomics.* **6**, 1656–1665.
53. Li, Y., Xu, X., Qi, D., Deng, C., Yang, P., Zhang, X. (2008) Novel Fe3O4@TiO2 core-shell microspheres for selective enrichment of phosphopeptides in phosphoproteome analysis. *J Proteome Res.* **7**, 2526–2538.
54. Xiong, S., Zhang, L., He, Q.Y. (2008) Fractionation of proteins by heparin chromatography *Methods Mol Biol.* **424**, 213–221.
55. Guerrier, L., Lomas, L., Boschetti, E. (2005) A simplified monobuffer multidimensional chromatography for high-throughput proteome fractionation. *J. Chromatogr.* **1073**, 25–33.
56. Fortis, F., Girot, P., Brieau, O., Boschetti, E., Castagna, A., Righetti, P.G. (2005) Isoelectric beads for proteome pre-fractionation. II: Experimental validation in a multicompartment electrolyzer. *Proteomics.* **3**, 629–638.
57. Restuccia, U. Boschetti, E., Fasoli, E., Fortis, F., Guerrier, L., Bachi, A., Kravchuk, A.V., Righetti, PG, Boschetti, E. (2009) pI-based fractionation of serum proteomes *versus* anion exchange after enhancement of low-abundance proteins by means of peptide libraries. *J Proteomics.* **72**, 1061–1070.

Chapter 3

Data Processing and Analysis Using ProteinChip® Data Manager Software

Enrique A. Dalmasso and Dominic Caseñas

Abstract

Mass spectrometry-based clinical proteomics and biomarker research require the processing of large numbers of patient samples in order to attain the statistical significance required to produce robust biomarker candidates. When processed using the high-sensitivity and high-throughput Surface Enhanced Laser Desorption/Ionization ProteinChip SELDI system, the result is an enormous amount of mass spectrometric profiling data. The time and effort required to mine this data and the quality of the candidate biomarkers generated is largely dependent on the quality and appropriate use of the software tools available. This chapter describes the typical workflow for processing and analyzing SELDI data using both univariate and unsupervised multivariate analysis tools.

Key words: Mass spectrometry, SELDI, ProteinChip, Data manager software

1. Introduction

The mass spectrometry (MS)-based techniques used in biomarker research generate vast quantities of data. For example, depending on the type of study and sample source, discovery phase experiments may require 100 or more samples and profile thousands of proteins and peptides, all with varying expression levels. With such high-dimensionality data, there is an inherent risk of missing out on strong markers in the face of so many other candidates, and a risk of finding markers that do not correlate with the clinical question being asked. Rigorous data analysis is required to sort through these data in order to select the most robust candidate biomarkers. The methods described in this chapter provide an overview of the data analysis workflow used for biomarker discovery with the

ProteinChip SELDI system and ProteinChip data manager software. For more detailed information, consult Bio-Rad's Biomarker Discovery Using SELDI Technology: A Guide to Data Processing and Analysis Using ProteinChip® Data Manager Software (1). A typical SELDI-based data analysis workflow involves the following steps:

- Data organization – creation or revision of sample annotations and organization of spectra into folders. Analysis of one condition requires organizing spectra from that condition into the same folder.

- Mass calibration – generation or updating of the calibration equation, which converts raw time-of-flight (TOF) data into mass data.

- Processing – baseline subtraction, filtering, noise estimation, alignment, and normalization of spectra. These steps optimize the accuracy and reproducibility of peak mass and intensity measurements prior to clustering.

- Generation of peak clusters – creation of peak clusters from spectral data, cluster editing, and preparation of peak cluster data for statistical analysis. Individual spectral features or peaks are labeled across all spectra collected under the same condition and then grouped together based on matches between their observed masses. The result is a list of masses, or peak clusters, that contain the same peaks across all spectra with associated intensities for each sample.

- Selection of candidate biomarkers – analysis of peak clusters by univariate statistical methods to identify candidate biomarkers that distinguish sample groups. The comparison of peak intensities of the same peak across multiple spectra (within a cluster) highlights differences in protein expression levels across multiple samples.

- Assessment of sample relationships – further evaluation of peak clusters using unsupervised multivariate statistics to reveal desired associations, examine possible sources of experimental variability (preanalytical or analytical bias), and generate more confidence in candidate biomarkers.

ProteinChip SELDI analysis begins with laser desorption and ionization of proteins from a ProteinChip array surface and detection by TOF-MS. Inside the ProteinChip SELDI reader, a nitrogen laser illuminates the sample, and the laser energy induces a change of state from the solid, crystalline state into the gas phase (desorption) with an associated ionization of the protein molecules. Once in the gas phase, the protein ions accelerate away from the metal array upon application of a voltage differential. The voltage differential imparts the same energy to all of the analytes in the sample, resulting in mass-dependent flight times. Once the protein ions strike the detector, the detector records the TOF of the analyte (which is then converted to a mass-to-charge ratio, or m/z) and the intensity of the response, which is directly related to the amount of that specific analyte on the array surface.

ProteinChip data manager software displays the MS data as a spectrum, plotting the measured signal intensity on the y-axis and the calculated m/z or mass on the x-axis. (Since the charge on most ionized proteins and peptides is 1, the m/z calculated from the TOF of a particle is equivalent to its molecular mass plus 1 Da.) The SELDI spectrum is the graphical representation of a sample's protein profile. Each spectrum has several notable features referred to in this chapter. They are:

- Matrix attenuation range – region in which most of the signal is generated by the matrix. This signal is suppressed by the instrument during data collection up to the designated attenuation setting to increase detection sensitivity. This region is excluded from further data processing and analysis.
- Low-mass range – region <20 kDa. Peaks in this range are generally sharp and well-resolved. Data collection and analysis parameters for this range are optimized to maximize peak resolution.
- High-mass range – region >20 kDa. Peaks in this range are generally of lower intensity and are broader due to chemical heterogeneity. Data collection and analysis parameters for this range are optimized to maximize detection sensitivity.

Within each spectrum, peaks represent the protein or peptide components of a sample, and their intensities correspond to the amounts of the species in a sample (their expression levels). Peaks are labeled either manually by the user or automatically by the software. In the manual method, the user applies the Centroid function of the software to click on and label peaks in the spectra. In the automatic method, the software identifies peaks by locating changes in intensity vs. the noise that are greater than a specified detection threshold. The software calculates peak intensity using the distance from the top of the peak to the calculated baseline. It calculates the mass by applying a calibration equation generated from a calibration sample (see Subheading 3.2 later in this chapter).

Though spectra and peaks may be two of the most recognizable forms of data generated by the ProteinChip SELDI system, peak clusters are the actual units of analysis used for biomarker discovery. Peak clusters are groups of peaks of similar mass that are treated as the same protein or peptide across multiple spectra. ProteinChip data manager software defines clusters within the spectra in each data folder. The data processing procedures described in this chapter are designed to optimize detection of the most robust clusters and selection of the most robust biomarker candidates. For biomarker data analysis, the normalized peak intensities from each spectrum within a cluster are used to compare expression levels for that substance between different sample groups.

Due to the inherent variability between patient samples, statistical methods are used for comparison of samples. Univariate statistical analyses compare only single features. In biomarker research, this means they consider differences in the intensity of a single peak at a time. The aim is to identify peak clusters with statistically significant differences in intensity between sample groups. Evaluating clusters using univariate statistics (using the P values and areas under receiver–operator curves, or AUC values) indicates their potential as single biomarkers. The final output from ProteinChip data manager software is a list of candidate biomarkers for further analysis. The next steps in the characterization, confirmation, and application of the biomarker candidates involve experimental validation, identification, and clinical assay implementation.

2. Materials

1. ProteinChip data manager software.
2. Spectra from multiple samples within a biomarker study, typically including replicates (duplicate or triplicate).
3. Relevant clinical, sample information, and processing information, including clinical parameters (such as patient age and sex, site of collection, tumor size, confounding factors, etc.), sample information (date and time of collection, sample classification, etc.) and sample processing information (fraction, array type, matrix, etc.).

3. Methods

3.1. Data Organization

The quality of data analysis relies not only on the quality of the spectral data but also on the quality of annotations and folder organization as well. ProteinChip data manager software stores spectral data in project folders and subfolders created during data collection and analysis. It uses the folder organization as the basis for subsequent peak cluster creation.

3.1.1. Ensure Proper Annotation of Spectra

Sample annotations are the descriptive information associated with a sample and its analysis. Annotations may include patient information (for example, patient name or sample number, disease group, disease staging or severity, drug treatment, age, and sex) as well as sample preparation details (for example, fraction number, matrix, stringency of binding, and wash buffers). Annotations are used for grouping spectra and as the basis for distinguishing sample groups during selection of biomarker candidates and development of classification algorithms. All data analyses rely on the information contained in sample annotations, so it is vital that all annotations are complete, correct, and unambiguous. In addition to unique identifiers, annotations should include clear, complete, and consistent descriptions of the samples and sample treatment procedures that were used.

1. Compile sample annotations prior to the start of experimental work.
2. If using ProteinChip data manager software, Enterprise Edition, use the Virtual Notebook feature to link sample annotations with the spectra.
3. Review the annotations to ensure they are complete before proceeding with data analysis.

3.1.2. Group Spectra into Folders

ProteinChip data manager software stores data in project folders and subfolders created during data collection and analysis. Spectra should be organized into separate folders within a project folder: base the organization on differences in experimental conditions, and separate analytical control and standard spectra from experimental spectra. ProteinChip data manager software allows you to easily organize spectra by selecting a series of fields for automatic segregation of spectra into conditions. ProteinChip data manager software generates clusters and performs biomarker analysis for all of the spectra within a folder. The list of biomarker candidates is specific for each experimental condition. To ensure comparison of relevant spectra and to avoid a potential source of analytical bias, be sure spectra are separated appropriately by condition.

1. Separate and group spectra according to the experimental conditions under which they were obtained, including fraction, array type, binding condition, matrix, and focus mass range.
2. Separate spectra collected from mass calibration standards and from QC and process control samples into different folders than those containing experimental spectra. Typical control samples include a reference sample used to assess reproducibility and blank spots for evaluating cross-contamination.

3.2. Mass Calibration

The ProteinChip SELDI system measures how long it takes for a molecule to travel the length of the flight tube and reports this time as TOF data. ProteinChip data manager software uses a calibration equation to convert TOF data into mass data in daltons. The mass calibration process involves two steps: internal calibration (generation of a calibration equation using standards of known mass) and external calibration (application of the calibration equation to spectra from experimental samples). Updating and applying the calibration equation regularly improves the mass accuracy of spectra.

3.2.1. Create the Calibration Equation Using Calibration Standards

Internal calibration generates the calibration equation using data obtained from mass calibration standards that contain peptides or proteins of known mass (for example, the ProteinChip all-in-one peptide and protein standards). The calibration equation is created by assigning known masses to corresponding peaks of the calibration spectrum. Internal calibration should generate a calibration equation with low residuals (errors), which are measured in parts per million (ppm) (see Note 1). Low residuals indicate better fit of the equation to the calibration data, and equations with low residuals are better suited to sample data. Using different sets of standards, multiple equations should be created for specific mass ranges to increase the accuracy within that range.

1. Select a calibration spectra that was collected using the same instrument settings and that contains standards that bracket the mass range of interest.
2. Manually label the calibrant peaks in a recently acquired peptide or protein calibration spectrum.
3. Use the Internal Calibration feature to create the calibration equation. Two types of curve fits are available: linear and quadratic. Linear is used when only three standards are present; quadratic should be used with four or more standards.
4. Evaluate the residuals of the calibration equation in the Internal Calibration dialog.

3.2.2. Improve Mass Accuracy Through External Calibration

The software automatically applies a default external calibration equation to all spectra during data collection. Update this equation on a regular basis to maximize the mass accuracy of the experimental data. External calibration should improve the mass accuracy of experimental spectra, which facilitates subsequent data analysis steps such as peak clustering. Properly calibrated data also enable comparison of biomarker candidates from different experimental conditions or across different studies.

1. Select an internally calibrated spectrum and use the Copy Calibration function to copy the calibration equation. Confirm that the internally calibrated spectrum was collected under the same conditions as the experimental data and that the calibration spectrum contains calibrants in the desired mass range.
2. Select all of the experimental spectra in a folder and use the Apply Calibration function.
3. Evaluate the effects of calibration.

3.3. Processing Spectral Data

Every data point generated by the mass spectrometer can theoretically be considered as a peak. Data processing is the step by which the software optimizes the spectrum to better differentiate true peaks (i.e. those derived from peptides or proteins) from the noise. During data collection, ProteinChip data manager software uses default parameters for processing spectral data. These processing steps include baseline subtraction, filtering, and noise estimation. The default processing parameters are a good starting point but may require further optimization for some sample types or conditions. This section describes how to optimize these processing parameters and reprocess the spectra prior to peak clustering. For optimization of processing parameters, the largest differences exist for data collected using different matrix molecules and mass optimization ranges. For each data folder, optimize the processing parameters for a few representative spectra, and then apply the changes to all other spectra in the folder.

3.3.1. Adjust Peak Intensity Calculations Using Baseline Subtraction

Mass spectrometers cannot always deliver a flat baseline. To compensate, ProteinChip data manager software applies a baseline subtraction algorithm to each spectrum. This algorithm subtracts the noise contributions to calculate more accurate peak intensities. The correct baseline of a spectrum should be relatively flat, and the peaks should be sharp and well resolved. Large humps in a spectrum indicate that the baseline subtraction algorithm may require adjustment. With the calculated baseline visible, the baseline should follow the base of all peaks in the spectrum. Gaps between the calculated baseline and spectrum indicate that the fitting width is too large; overlaps indicate that the fitting width is too small.

1. Examine the shape of the baseline in several reference or representative spectra in each folder. Toggle the Subtract Baseline function to visualize the baseline.
2. If necessary, adjust the values for the Fitting Width and Smoothing settings.
3. Evaluate effects of any change on the appearance of the spectrum, and work to generate a baseline that is consistent with the shape of the spectrum.
4. For optimum reproducibility, apply the same baseline subtraction settings to all spectra within a condition.

3.3.2. Reduce Noise by Filter Adjustment

Filtering removes noise from a spectrum to improve the signal-to-noise ratio, particularly in the high-mass range, thereby improving peak detection. Changing the filtering setting is optional; the default settings are effective for most situations. The objective of filtering is to decrease the amount of noise in the spectrum, but overfiltering may result in loss of resolution. The amount of filtering required for a spectrum depends mainly on the mass range being analyzed.

1. Specify the value for noise filtering. In most cases, the default Average filter with a setting of 0.2 times expected peak width is appropriate. For higher masses (10–200 kDa) a wider setting (typically 1.0) can be used.

2. Analyze the effects of any changes to the filter settings on the spectrum. If resolution is compromised, revert to the original settings.

3. Apply the same filter setting to all spectra in a condition.

3.3.3. Improve the Signal-to-Noise Ratio by Setting the Noise Range

The noise of a spectrum represents the average intensity of the data points in areas where there are no peaks. Noise is a measure of the variation in the signal, not of the signal itself. Adjusting the noise calculation improves signal-to-noise ratios, thereby improving peak detection. It is important to set an appropriate range for defining the spectrum noise (for example by excluding the signal generated by the matrix or by matrix attenuation) in order to optimize the sensitivity and reproducibility of automatic peak detection.

1. Set the starting mass for noise calculation to the matrix attenuation value (specified during data collection) plus ≥10%.

2. Analyze any effects of this change on the peak counts from automatic peak detection (no visible change is made to the spectrum).

3. Apply the optimized noise setting to all spectra in a condition.

3.4. Normalize and Align Spectra

Perform normalization and alignment for all the spectra within each folder (condition) at the same time. Normalization and alignment help to maximize reproducibility and mass precision enabling selection of the most robust peak clusters. This section describes how to perform normalization and alignment.

3.4.1. Standardize Intensities by Normalization

Total ion current (TIC) normalization standardizes the intensities of a set of spectra to compensate for any spectrum-to-spectrum variations. Variations may be caused by minor differences in total protein concentration, sample preparation, or data collection. Generally, all spectra in the condition should be selected for TIC normalization at one time. Proper normalization improves spectrum-to-spectrum reproducibility. Normalization also generates normalization factors, which can be used to identify spectra of poor quality.

1. Use the Normalize Spectra function to normalize all spectra within a condition by their TIC. Specify the minimum starting mass, the ending mass, and whether to use an external normalization coefficient.

2. Generate a histogram displaying normalization factors. Inspect spectra with normalization factors outside the normal distribution

to assess the protein profiles, and eliminate spectra of poor quality prior to statistical analysis.

3. If outlying spectra are removed from the analysis, repeat step 1 (not required if normalizing to an external coefficient).

3.4.2. Improve Mass Precision by Spectrum Alignment (Optional)

Spectrum alignment reduces mass variations of peaks within a group of similar spectra. After alignment, the same peaks across all spectra have higher mass precision. This step is optional and often unnecessary for data collected using instruments that have been properly calibrated. It is recommended when clustering is problematic, for example when a shoulder is labeled instead of a peak apex. The goal of alignment is to increase the mass precision of peaks across all spectra within a condition. Following alignment, aligned peaks are plotted in the Alignment results window. The window also shows the pooled %CV for mass, which should be lower following spectrum alignment.

1. If mass shifts occur between spots or arrays, label at least five well-resolved peaks in the mass range of interest using the Centroid All tool to ensure inclusion of all spectra. Ideally all peaks used for alignment should be present in all spectra, but the alignment algorithm can tolerate some gaps. For more variable data with gaps, use a larger number of peaks for alignment and ensure that no spectrum is lacking most of the peaks used for alignment.

2. Select the reference spectrum, which serves as the standard mass template.

3. Evaluate the effect of alignment, and either save the results or revert back to the default setting if the alignment was ineffective. If alignment is not effective, confirm that all peaks are labeled correctly and label additional peaks to improve alignment.

3.5. Generation of Peak Clusters

Peak clusters are groups of peaks of similar mass that are treated as the same substance (protein or peptide, for example) across multiple spectra. ProteinChip data manager software uses statistical analyses of peak clusters to uncover biomarker candidates that exhibit different expression levels between sample groups. This section describes how to create and edit peak clusters.

3.5.1. Associate Peaks from Multiple Spectra into Peak Clusters

Peak clustering labels peaks across all the spectra in a folder and then groups them together based on matches between their calculated masses. The result is a list of masses, or peak clusters, that contain peaks of similar mass from each spectrum with associated intensities for each sample. This list is interpreted as a list of peptides and proteins with their associated expression levels. Ideally, peak clusters contain all the peaks from the spectra within a condition, with each spectrum contributing one peak to each cluster. If peaks are absent from a spectrum, baseline intensities are estimated and included for statistical calculations. The cluster table lists all the clusters and provides statistics (such as minimum and maximum values and %CV) for intensity and mass.

1. Ensure that all data processing steps have been performed and that all annotations have been added or updated prior to clustering.
2. Delete all existing peak labels.
3. Create peak clusters using the Create EDM (expression difference mapping) feature by specifying parameters for peak detection and cluster completion.

3.5.2. Adjust Peaks Within Clusters Using Cluster Editing

To ensure accurate statistical analysis, the output of peak clustering must be complete and precise. Mislabeling of peaks may occur if peaks are very broad, contain shoulders, or if the mass window for peak clustering was too wide. Peak labels that are not automatically placed at the appropriate peak centroid can be manually adjusted to the correct location using the cluster editing process within ProteinChip data manager software. This allows you to shift peak labels within a cluster, without having to repeat the cluster generation process for the entire condition. Cluster editing resolves any problems caused by mislabeling of peaks and results in more accurate statistical analysis of biomarker candidates.

1. Use the Plot Spectra in Cluster function from the Clusters tab to view spectra.
2. Examine the spectra to identify peak labels that are not placed appropriately, as compared to other spectra in that peak cluster.
3. Use the Edit Cluster or Move Peak(s) cursor to move a peak label from one position on the spectrum to another.
4. Update the statistics in the cluster table.

3.6. Selection of Candidate Biomarkers

Statistical analyses of peak clusters are used to select clusters that exhibit significant differences in intensity values (expression levels) between relevant sample groups. In biomarker research, univariate statistical methods are used to consider differences in the expression levels of one biomarker candidate at a time. This section describes the selection and evaluation of single-candidate biomarkers using univariate analyses.

3.6.1. Screening for Biomarker Candidates Using P Values

In ProteinChip data manager software, the P value calculation rapidly screens peak cluster data for biomarker candidates using the Mann-Whitney or Kruskal-Wallis tests. After locating biomarker candidates by sorting for low *P* values (see Note 2), assess the quality of the underlying peaks using the Plot Spectra in Cluster feature. In this step, differentiate clusters consisting of high-quality peaks from low-quality clusters containing noisy or spurious peaks. Exclude low-quality clusters from further analysis, since the resulting peak intensities are less reliable and more likely to yield false discoveries.

1. Use the Calculate *P*-values function to specify the grouping criterion that differentiates the experimental groups within a condition.
2. Average replicates by specifying the field that contains the replicate information (usually the Sample name field).
3. Examine the *P*-value column for peak clusters that display statistically significant differences between groups.
4. Assess the quality of the underlying peaks using the Plot Spectra in Cluster feature. Manually inspect peak clusters with good *P* values for peak quality (good signal-to-noise ratio, resolution, accurate peak labeling, etc.) using the Spectrum Viewer and the three scatter plots.

3.6.2. Measure Candidate Biomarker Sensitivity and Specificity Using ROC Curves

For two-group comparisons, use ROC curves to compare the sensitivity and specificity of a biomarker candidate at different cutoff values for peak intensity. The sensitivity and specificity for a group (for example, control vs. treated or normal vs. disease) is related to the amount of overlap in protein expression levels, and this overlap is quantifiable using the area under the ROC curve (AUC). The ROC curve plots sensitivity vs. 1-specificity and provides an evaluation of the ability of a biomarker candidate to indicate a disease, response to treatment, or other difference in sample groups. The area under the plot, the AUC, provides a quantitative measurement of this potential: as the AUC approaches a value of 1.0, it indicates higher sensitivity and specificity, whereas an AUC of 0.5 indicates no separation (see Note 3).

1. Use the *P*-Value Wizard to automatically construct ROC plots and calculate AUC values.
2. Examine the ROC area column of the cluster table to identify peak clusters with good AUC values.

3.7. Assessment of Sample Relationships

In addition to univariate analyses, unsupervised multivariate analyses such as hierarchical clustering and principal component analysis (PCA) can be used to reveal both desired (disease diagnosis, drug treatment efficacy, adverse drug effect) and undesired (preanalytical or analytical bias) associations within the data. These unsupervised techniques ignore the sample groupings provided and find inherent differences or separation in the data set based on peak intensities, without bias from the properties of the samples that generated them. This section describes the application of hierarchical clustering and PCA for examining the entire protein expression profile and revealing sample relationships within an experiment.

3.7.1. Examine Sample Hierarchical Clustering Patterns

Hierarchical clustering is an unsupervised, exploratory learning method of multivariate analysis that groups multidimensional data by similarity. This method generates heat maps and dendrograms, widely used visualizations of high-dimensional data such as microarray data in gene expression studies. The objective of unsupervised hierarchical clustering is to observe groupings of spectra based on their complete expression profiles.

In the clustered heat map, the arrangement of samples can indicate accurate clinical grouping, unknown confounding factors, or possible subgroups with common clinical characteristics (for example, cancer metastasis vs. no metastasis subgroups may generate two separate subclusters within the group previously defined as cancer). The proximity of spectra to each other and the branching patterns in the dendrogram reflect sample similarity. Spectra that are in close proximity within a branch exhibit similar expression profiles: the branch length reflects the degree of the relationship, with shorter branches indicating closer relationships. Options allow you to display the relationships and similarities between expression profiles of samples (columns) and between expression levels of peak clusters (rows).

1. Select the sample group for examination using the P-Value Wizard.
2. Launch the Heat Map feature to visualize peak clusters as green and red points with variable shades based on relative expression level.
3. Use the dendrogram to create a clustered heat map that illustrates sample relationships.
4. Examine the clustered heat map for correlations between patterns of samples and the grouping criterion.

3.7.2. Visualize Sample Similarity Using PCA

PCA is another unsupervised multivariate analysis method used to visualize multidimensional data by mathematically reducing dimensionality of data into components that approximate the original multidimensional data. In the case of ProteinChip SELDI data, each cluster of peak intensities is considered an independent dimension. ProteinChip data manager software uses PCA to visualize spectra in two- and three-dimensional graphs that illustrate relationships between spectra based on their expression profiles.

The objective of unsupervised PCA is to visualize the degree to which the samples in the experiment are similar. Unlike other visualizations in ProteinChip data manager software, where a dot on a plot represents a peak, each dot on a PCA plot represents the expression profile of an entire spectrum. Simply put, the closer dots on a PCA plot, the more similar the expression profiles. If the colors representing a particular sample characteristic (i.e. sample group, collection site, bioprocessor) appear close together and separate from others in a PCA plot, those samples are closely related and exhibit differences from other sample groups. If the colors are interspersed, there is no separation of bias based on that sample characteristic.

1. Select a grouping criterion using the *P*-Value Wizard.
2. Launch the PCA feature to automatically generate PCA plots.
3. Examine the distribution of the samples in the PCA plots to identify possible associations with the grouping criterion.

4. Notes

1. With the ProteinChip all-in-one peptide standard (Bio-Rad catalog #C10-00005), errors of <300 ppm are acceptable and errors of <100 ppm are achievable. With the ProteinChip all-in-one protein standard II (Bio-Rad catalog #C10-00007), errors of <1,000 ppm are acceptable. If high residuals are observed, adjust the peaks chosen as calibrants and ensure that:
   (a) Appropriate peaks were selected.
   (b) Peaks were assigned the appropriate mass.
   (c) All calibrant peaks were labeled at the apex (zoom in on each peak during labeling).
2. No formal *P* value threshold applies for all experiments. The numbers of samples and candidate biomarkers, as well as other statistical principles, such as false discovery rate, all play a major role in determining the appropriate *P* value for a given experiment. However:
   (a) *P* values of <0.05 are generally considered significant.
   (b) *P* values of <0.01 are a reasonable starting point for more robust markers.
3. AUC is calculated with respect to one of the groups, which is called the positive group. Since this designation is arbitrary in a biomarker discovery experiment, the AUC for a good biomarker

candidate may also be close to 0.0. Changing the positive group inverts the value to calculate the true AUC.

(a) As AUC values approach 1.0 and 0.0, the ability of that biomarker candidate to separate relevant grouping improves. Typically, values of >0.7–0.8 are considered significant.

(b) A value of 0.5 indicates that the protein expression levels of both groups completely overlap.

Reference

1. Biomarker Discovery Using SELDI Technology: A Guide to Data Processing and Analysis Using ProteinChip® Data Manager Software, Bio-Rad Bulletin # 5814.

Chapter 4

Purification and Identification of Candidate Biomarkers Discovered Using SELDI-TOF MS

Amanda L. Bulman and Enrique A. Dalmasso

Abstract

Purification and identification of candidate biomarkers is a critical step in the biomarker development process, since it provides insight into the disease biology and facilitates the development of analyte-specific assays. Top-down biomarker discovery workflows like SELDI-TOF MS yield candidate markers that are identified based on native mass. Positive identification of these candidate biomarkers requires further enrichment and/or purification. While purification methods must be optimized for each protein target, there are two general workflows. Native peptides under approximately 4 kDa can be subjected to direct sequence analysis using a tandem mass spectrometer whereas proteins over approximately 4 kDa usually require proteolytic digestion prior to MS/MS analysis. In both cases, partial purification is usually necessary to enrich the candidate biomarker relative to other proteins in a complex biological mixture. This chapter provides detailed protocols for protein purification (including anion exchange, metal affinity, and reverse phase chromatography as well as SDS-PAGE) and identification (including protein processing, digestion, and database searching).

Key words: Protein purification, Tandem mass spectrometry, Protein identification, SELDI, Proteolytic digestion, Anion exchange chromatography, Immobilized metal affinity chromatography, Reverse phase chromatography

1. Introduction

While top-down proteomics methods allow detection of post-translation modification and rapid analysis of statistically relevant numbers of samples, positive identification usually requires partial purification and/or enrichment of the target biomarker. There are two different paths to positive identification, depending on the molecular weight of the target biomarker (Fig. 1). Peptides under approximately 3–4 kDa can often be sequenced directly, without the need for proteolytic digestion (1–4). The upper mass limit for

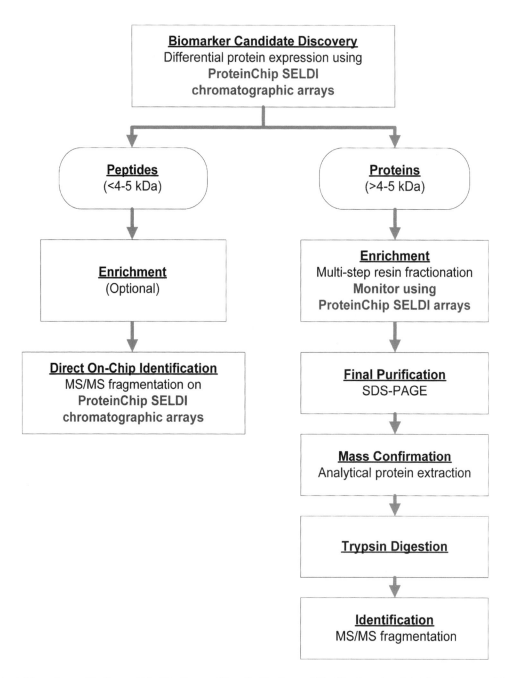

Fig. 1. Biomarker purification and identification workflow. Purification and identification of protein biomarkers can follow one of two paths, depending primarily on the molecular weight of the target biomarker. Peptides under approximately 4 kDa can be analyzed directly on a tandem mass spectrometer without proteolytic digestion. While some peptides can be sequenced directly from profiling arrays, most will require one or more purification steps to enrich the target and reduce the sample complexity. Proteins over approximately 4 kDa require proteolytic digestion prior to sequence analysis. These targets generally require two or more chromatographic purification steps, followed by SDS-PAGE. The objective of the purification steps is to enrich the target protein relative to other proteins of similar size such that the gel band containing the target protein can be distinguished from other neighboring protein bands. Intact proteins are eluted from the gel to confirm native mass and purity before performing proteolytic digestion and MS/MS analysis.

direct ID of a peptide in a complex mixture depends, in part, on the MS instrumentation being used. LC-MS approaches typically have a practical upper limit of approximately 3 kDa, whereas some MALDI TOF-TOF systems are capable of extending this limit to 4–5 kDa. Some peptide biomarkers can be sequenced directly from the profiling arrays, while others will require one or more enrichment steps prior to identification.

Peptides and proteins that are too large for direct sequence analysis (typically >4 kDa) are usually subjected to one or more chromatographic purification steps in order to enrich the target protein relative to other proteins of similar size (4–12). The final purification step is usually performed using SDS-PAGE and selected bands excised and subjected to proteolytic digestion. Proteolytic digestion yields peptides of a sufficiently small size for MS/MS analysis. Positive identification of larger proteins typically requires identification of two or more peptides from the digest.

To identify a candidate biomarker using either of these two approaches, a sufficient amount of the target protein (~1 pmol) must first be enriched. Many different protein purification and enrichment techniques are available, which take advantage of unique physiochemical properties. Since each protein is unique, some degree of trial and error is required to optimize purification conditions for a given target. However, while the goal of many standard protein purification workflows is to achieve greater than 95% purity, much lower purity can be tolerated for protein identification. To facilitate MS-based protein identification, the target peptide or protein must be sufficiently enriched to generate high quality MS/MS spectra (for peptides under ~4 kDa) or to clearly separate from proteins of similar mass on an SDS-PAGE gel (for proteins and peptides over ~4 kDa). While a wide variety of other purification techniques are available (i.e. precipitation, filtration, etc.), for SELDI-based biomarkers, the emphasis is usually on chromatography.

The SELDI-TOF biomarker discovery workflow utilizes chromatographic separation to reduce sample complexity and visualize more total protein species. Therefore, the behavior of a given protein during the discovery phase study reveals information about a protein's physiochemical properties that can be extremely valuable in developing a purification scheme. During discovery, anion exchange and ProteoMiner enrichment steps are frequently used as pre-fractionation steps to reduce the concentration of high abundance proteins, thereby increasing the total number of proteins detected (13–16). Further separation and enrichment are achieved by binding fractions (or crude samples) to multiple ProteinChip array chemistries. For example, fractions from ProteinMiner enrichment are typically profiled on four different array chemistries, including reverse phase (H50), metal affinity (IMAC30), and anion (Q10) and cation (CM10) exchange. The binding (and non-binding) characteristics of the target protein during pre-fractionation and profiling on ProteinChip arrays provides valuable information on separation techniques that may be useful in enriching the target protein and/or removing other proteins of similar mass.

In some cases, a single step-enrichment step is sufficient for identification, and in other cases, multidimensional strategies are required. In cases where pre-fractionation was used during the discovery phase study, the same technique is typically scaled up and used as the first step of purification. While there are a wide range of methods that can be used for subsequent purification, we have found that three types of chromatography often provide sufficient separation, particularly when orthogonal chemistries are combined in a multidimensional purification scheme. The three most commonly used chromatographic methods are anion exchange, metal affinity (using copper ions) and reverse phase chromatography and are described below. These can be used as single-step, two-step, or three-step purification.

- Anion exchange chromatography is often used as the first step of purification and exploits charged amino acid side chains for separation of proteins by pI. Binding is normally performed at high pH (pH 9.0) to ensure binding of most proteins, and proteins are subsequently eluted either using decreasing pH or increasing salt.
- Immobilized metal affinity chromatography (IMAC) is often used in series with anion exchange chromatography but can also be used as the first step of purification as an alternative to anion exchange chromatography. Copper is usually used as the metal ion. Binding (through histidine side chains) is normally conducted at physiological pH in the presence of salt, which reduces non-specific ionic interactions. Proteins can be eluted with increased concentrations of imidazole, which acts as a competitor for binding to copper ions.
- Reverse phase chromatography is typically used as the final purification step prior to SDS-PAGE and/or direct sequence analysis (for peptides <4 kDa). While this step provides some separation by size and hydrophobicity, the most important function of this step is often to desalt and concentrate samples prior to SDS-PAGE or MS/MS analysis. A single reverse phase enrichment step is sometimes sufficient for enrichment of native peptides for direct sequence analysis.

It is critical to monitor the protein of interest throughout the purification process, both to assess purity and to confirm the presence of the target protein. Several comparisons are used to facilitate tracking of the target protein, including (1) processing samples with high and low concentrations of the target protein in parallel at all steps, (2) monitoring fractions for the presence of the target protein by profiling on the same array chromatographic chemistry used during discovery, and (3) monitoring overall purity on normal phase arrays.

- When sample volumes are limited, individual samples can be pooled to create two pools. While pooling is never recommended during biomarker discovery due to its negative impact on statistical power, pooling is a viable alternative during purification. When selecting individual samples to pool for purification consider both the intensity of the target protein and the relative intensity of other peaks of similar mass that will need to be removed during purification.

- Throughout purification, fractions should be monitored on both chromatographic and normal phase ProteinChip arrays. The chromatographic surface on which the candidate marker was originally discovered, combined with the protein mass, serves as the assay throughout purification. Since many proteins do not bind to all chromatographic surfaces, fractions may also be monitored on normal phase (NP20) arrays to assess overall purity. When using NP20 arrays, samples with high salt concentrations (i.e. flow through fractions) will usually require a water wash to remove salts.

The development of a purification/enrichment scheme for a given target protein is an iterative process. While the number of steps and the combinations of chromatographic chemistries required for purification will vary for different targets, thorough analysis of all available data can be used to effectively guide selection of appropriate methods. Where a multidimensional approach is required, chromatographic media should be selected to best utilize orthogonal separation chemistries. This chapter provides detailed methods for protein purification using anion exchange, IMAC, and reverse phase chromatography. Protocols required for processing (SDS-PAGE, reduction and alkylation, and proteolytic digestion) and guidelines for positive identification (MS/MS collection and database searching) are also provided. These methods have been used successfully to purify and identify protein biomarkers from serum (1, 2, 6–8), cerebrospinal fluid (4, 9), urine (5, 11), cell and tissue lysates (10), cell culture media (12), and other biological fluids (3).

2. Materials

2.1. Equipment and General Supplies

1. ProteinChip SELDI system or Lucid Proteomics package.
2. Tandem mass spectrometer.
3. Swinging bucket centrifuge.

4. Microcentrifuge.
5. Microplate agitator (recommend DPC MicroMix 5).
6. SpeedVac.
7. Vortex mixer.
8. ProteinChip cassette-compatible bioprocessor (Bio-Rad).
9. ProteinChip arrays (Bio-Rad).
   (a) NP20 normal phase arrays.
   (b) Array chemistries on which target protein was originally discovered (typically CM10, H50 hydrophobic, IMAC30, and/or Q10 arrays).
10. Plastic consumables (see Note 1).
    (a) Eppendorf safe-lock microcentrifuge tubes (0.6, 1.5, and 2.0 ml).
    (b) Eppendorf standard pipette tips and gel-loading tips.
    (c) Bio-Spin (1.2 ml) and/or Mini Bio-Spin (0.8 ml) spin columns.

2.2. Anion Exchange Chromatography

1. Anion exchange: Q ceramic HyperD F (Pall Corp.).
2. Rehydration buffer: 50 mM Tris–HCl, pH 9.0.
3. U9 buffer: 9 M urea, 2% CHAPS, 50 mM Tris–HCl, pH 9.0.
4. U1 buffer (U9 buffer diluted 8:1 with rehydration buffer): 1 M urea, 0.22% CHAPS, 50 mM Tris–HCl, pH 9.0.
5. Elution buffers.
   (a) pH 9 = 50 mM Tris–HCl, 0.1% octyl β-D-glucopyranoside (OGP), pH 9.0.
   (b) pH 7 = 50 mM HEPES with 0.1% OGP, pH 7.0.
   (c) pH 5 = 100 mM sodium acetate with 0.1% OGP, pH 5.0.
   (d) pH 4 = 100 mM sodium acetate with 0.1% OGP, pH 4.0.
   (e) pH 3 = 50 mM sodium citrate with 0.1% OGP pH, 3.0.
   (f) Organic = 33.3% isopropanol (IPA)/16.7% acetonitrile/0.1% TFA.

2.3. Immobilized Metal Affinity Chromatography

1. IMAC HyperCel resin (Pall Corp.).
2. IMAC charging solution: 100 mM $CuSO_4$.
3. IMAC neutralization solution: 100 mM sodium acetate, pH 4.0.
4. IMAC binding buffer: 100 mM phosphate buffer, pH 7.2, 0.5 M NaCl.
5. Elution buffers:
   (a) 10 mM imidazole in IMAC binding buffer.
   (b) 20 mM imidazole in IMAC binding buffer.

(c) 30 mM imidazole in IMAC binding buffer.
(d) 40 mM imidazole in IMAC binding buffer.
(e) 50 mM imidazole in IMAC binding buffer.
(f) 100 mM imidazole in IMAC binding buffer.
(g) 200 mM imidazole in IMAC binding buffer.

2.4. Reverse Phase Chromatography

1. Reverse phase support: PLRP-S 300A 10–15 µM (Varian Inc.).
2. 50% acetonitrile, 5% trifluoroacetic acid (TFA).
3. Elution buffers:
   (a) 10% acetonitrile, 0.1% TFA.
   (b) 20% acetonitrile, 0.1% TFA.
   (c) 30% acetonitrile, 0.1% TFA.
   (d) 40% acetonitrile, 0.1% TFA.
   (e) 50% acetonitrile, 0.1% TFA.
   (f) 75% acetonitrile, 0.1% TFA.

3. Methods

3.1. Purification

3.1.1. Anion Exchange Chromatography

Anion exchange chromatography is often used as the first step of purification, especially when anion exchange fractionation has been used during discovery to reduce sample complexity. Anion exchange fractionation is compatible with many sample types, but may not be ideal for samples with high concentrations of salt (i.e. urine, some lysates). Example spectra of anion exchange fractionation of an *E. coli* extract are shown in Fig. 2. Since many proteins associate with each other in plasma, urea and detergent are typically added to the samples prior to fractionation to reduce protein–protein interactions. Binding is usually conducted at a high pH (pH 9.0) and proteins eluted in a stepwise manner using buffers of decreasing pH. Increasing salt concentrations may also be used as an alternative to the pH gradient.

1. Equilibrate Q HyperD® F resin (BioSepra) by adding approximately 4 resin volumes of 50 mM Tris–HCl, pH 9.0. Mix gently, remove buffer, and repeat three times.
2. Add resin to a spin column.
3. Equilibrate with approximately 2 column volumes of U1 buffer. Repeat.
4. Thaw samples on ice.
5. Mix sample with 1.5× the same volume of U9 buffer and incubate 30 min at room temperature with shaking.

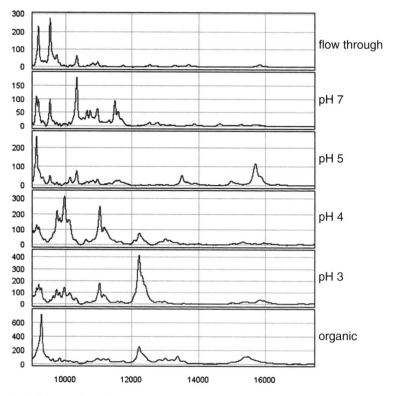

Fig. 2. Anion exchange fractionation. Anion exchange chromatography was used as the first enrichment step in the purification of proteins from *E. coli*. The *E. coli* lysate was adjusted to pH 9.0 and applied to Q HyperD F resin. The flow through was collected and bound proteins eluted with decreasing pH.

6. Dilute sample with an equal volume of U1 buffer prior to loading on the column.

7. Apply sample to an anion exchange column (see Note 2) and mix for 30 min.

8. Centrifuge the column(s) at $1,000 \times g$ for 1 min and label the eluate as Flow Through.

9. Add 500 µl of 50 mM Tris–HCl (pH 9), 0.1% OGP to the column and mix gently for 10 min.

10. Centrifuge the column(s) at $1,000 \times g$ for 1 min and label the eluate as the pH 9 fraction (or fraction 1).

11. Repeat steps 8 and 9 using the following elution buffers and labeling schemes
    (a) 50 mM HEPES (pH 7), 0.1% OGP – pH 7 fraction (or fraction 2).
    (b) 100 mM sodium acetate (pH 5), 0.1% OGP – pH 5 fraction (or fraction 3).
    (c) 100 mM sodium acetate (pH 4), 0.1% OGP – pH 4 fraction (or fraction 4).

(d) 50 mM sodium citrate (pH 3), 0.1% OGP – pH 3 fraction (or fraction 5).

(e) 33% IPA/17% ACN/0.1% TFA solution – organic fraction (or fraction 6).

12. Analyze fractions on the array chemistry on which the target was originally discovered to confirm the presence and relative intensity of the target protein in the positive vs. negative samples. Use 5–10 µl of each fraction, diluted tenfold in the appropriate binding buffer.

13. Analyze 1–2 µl of each fraction on NP20 arrays to assess overall purity and presence of protein "contaminants" of similar mass (see Note 3).

3.1.2. Immobilized Metal Affinity Chromatography

Metal affinity chromatography can be used as a first or second step of purification. While IMAC resins can be charged with many different divalent or trivalent metals, Cu^{2+} is most frequently used for purification. IMAC chromatography is tolerant of salts and may therefore be a better choice than anion exchange for samples containing high salt concentrations. Proteins are typically bound at neutral pH and eluted with increasing concentrations of imidazole (Fig. 3).

1. Charge IMAC-HyperCel resin with 2–3 volumes of 100 mM copper sulfate.

2. Neutralize with Neutralization Buffer.

3. Wash with proteomics-grade water.

4. Equilibrate resin with 5 resin volumes of IMAC binding buffer.

5. Remove equilibration buffer and add sample, adjusted to ~pH 7, 0.5 M NaCl. Incubate a minimum of 1 h at room temperature on an end-over-end rotator. Samples may also be incubated overnight at 4°C.

6. Centrifuge the column(s) at $1,000 \times g$ for 1 min and label the eluate as Flow Through.

7. Add 1.5 resin volumes of 5 mM imidazole in IMAC binding buffer and mix gently for 10 min.

8. Centrifuge the column(s) at $1,000 \times g$ for 1 min and label the eluate as the 5 mM imidazole fraction.

9. Repeat steps 6 and 7 using the following elution buffers with increasing imidazole concentrations:

    (a) 10 mM imidazole in IMAC binding buffer.

    (b) 20 mM imidazole in IMAC binding buffer.

    (c) 30 mM imidazole in IMAC binding buffer.

    (d) 40 mM imidazole in IMAC binding buffer.

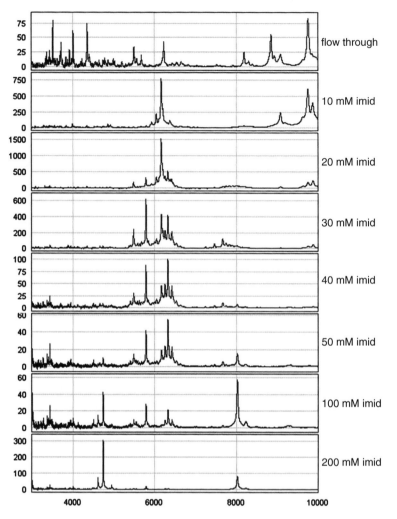

Fig. 3. IMAC chromatography. While IMAC chromatography is often used following anion exchange, it can also be used as the first purification step, especially in samples with high salt concentrations, which interfere with binding to anion exchange. In this example, human urine was diluted in IMAC buffer and bound to IMAC resin. Bound proteins were eluted with increasing concentrations of imidazole.

(e) 50 mM imidazole in IMAC binding buffer.
(f) 100 mM imidazole in IMAC binding buffer.
(g) 200 mM imidazole in IMAC binding buffer.

10. Analyze fractions on the array chemistry on which the target was originally discovered to confirm the presence and relative intensity of the target protein in the positive vs. negative samples. Use 5–10 µl of each fraction, diluted tenfold in the appropriate binding buffer.

11. Analyze 1–2 µl of each fraction on NP20 arrays to assess overall purity and presence of contaminants of similar mass.

3.1.3. Reverse Phase Chromatography

Reverse phase chromatography is typically used as the last step in purification, In addition to providing separation based on hydrophobicity, the reverse phase step provides desalting and concentration, thereby preparing the sample for either direct MS/MS analysis (for peptides <4 kDa) or gel electrophoresis (for targets >4 kDa). Binding is conducted under conditions that would result in all proteins binding to the resin, and proteins are eluted using increasing concentrations of acetonitrile (Fig. 4).

1. Equilibrate PLRP-S beads (Varian) with 10% ACN/0.1% TFA. The amount of reverse phase resin required can vary depending on the total protein, but 50 μl of resin is usually sufficient for most purified preparations.
2. Adjust samples to a final concentration of 5% ACN/0.5% TFA.
3. Add sample to a tube containing PLRP-S beads (see Note 4) and mix for a minimum of 30 min at room temperature. Samples can also be incubated at 4°C overnight.

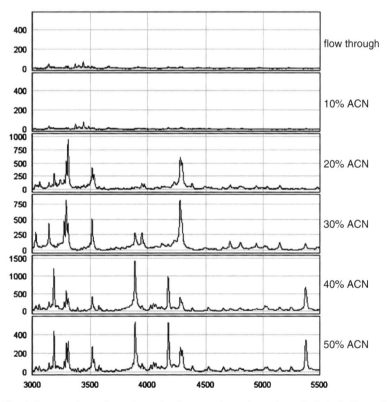

Fig. 4. Reverse phase chromatography. Reverse phase chromatography is typically used as the final column chromatography step, as it concentrates and desalts the sample, preparing it for direct sequence analysis or SDS-PAGE. In the example shown here, human serum was first subjected to anion exchange fractionation. The flow through was then applied to IMAC-Cu resin. The 40 mM imidazole eluate from the IMAC step was then applied to reverse phase beads and eluted with increasing concentrations of acetonitrile.

4. Centrifuge the tube at 800×g for 1 min, carefully remove the supernatant by aspiration, and label as FlowThrough (see Note 5).

5. Add approximately 2 resin volumes of 10% acetonitrile in 0.1% TFA and incubate 10 min at room temperature.

6. Centrifuge the tube at 800×g for 1 min, remove the supernatant by aspiration, and label as 10% ACN fraction (or fraction 1).

7. Repeat steps 5 and 6 using the following elution buffers with increasing acetonitrile concentrations:
   (a) 20% acetonitrile, 0.1% TFA.
   (b) 30% acetonitrile, 0.1% TFA.
   (c) 40% acetonitrile, 0.1% TFA.
   (d) 50% acetonitrile, 0.1% TFA.
   (e) 75% acetonitrile, 0.1% TFA.

8. Analyze 1–2 µl of each fraction on NP20 arrays to assess overall purity. Note that the flow through and 10% ACN fractions should contain little or no protein. Significant protein in these fractions indicates that the resin was overloaded. If the target protein is detected in the flow through, bind the flow through to fresh resin and repeat steps 4–7e.

3.1.4. Gel Electrophoresis

1. Concentrate samples in a Speed-Vac.
2. Re-dissolve in 20 µl of the appropriate gel loading buffer.
3. Analyze by SDS-PAGE. Gels and protein standards should be selected to maximize separation in the mass range of the protein target (see Note 6).
4. Stain and destain gel according to manufacturer's instructions.

3.2. Identification of Proteins Greater Than ~4 kDa

3.2.1. Analytical Extraction

Analytical extraction is critical to ensuring that the correct protein is identified. Since many proteins do not run true to size using SDS-PAGE, it is often difficult to accurately predict which band contains the protein of interest and the purity of the target protein in the band. Analytical extraction allows confirmation of the presence of the target protein and a more accurate assessment of purity.

1. Excise gel bands of interest from the polyacrylamide gel with a clean razorblade or a 1-mm diameter Harris micro-punch (Sigma). If using a razorblade, cut band into four smaller pieces.

2. Place 1/4 of the gel pieces into a 0.5-ml microfuge tube.

3. Wash gel pieces with 100 µl of 50% methanol/10% acetic acid for 30 min.

4. Remove solvent (see Note 7) and dehydrate with 50 µl of ACN for 15 min. Remove solvent.

5. Extract intact proteins from the gel by adding 10 µl of 50% formic acid/25% ACN/15% IPA and incubating for 3 h at room

temperature with vigorous shaking. If resulting spectra are weak (see Step 6), the incubation can be extended overnight.

6. Profile 2 µl of the extract on an NP20 ProteinChip array.

3.2.2. De-Staining

1. After excising the band of interest (see Subheading 3.2.1, step 1), place the remaining 3/4 of the gel pieces into a 0.5-ml microfuge tube.
2. Add 400 µl 50% methanol, 10% acetic acid and shake vigorously for 1 h.
3. Remove the solvent, taking care not to remove the gel pieces, and repeat step 1, shaking for 30 min.
4. Remove the solvent and add 400 µl 50% acetonitrile, 100 mM ammonium bicarbonate, pH 8. Shake for 1 h.
5. Remove the solvent and add 100 µl 100% acetonitrile. Shake for 15 min or until gel pieces turn white.
6. Remove acetonitrile and dry the gel pieces in a vacuum concentrator (see Note 8).

3.2.3. In-Gel Reduction and Alkylation

Reduction and alkylation is an optional step but can improve sequence coverage by allowing positive identification of peptides containing disulfides. Reduction and alkylation of proteins over 4–5 kDa is normally conducted in-gel following SDS-PAGE, which is described in the section below. Reduction and alkylation can also be conducted in solution, which is described in Subheading 3.3.1.

1. Add 100 µl 5 mM DTT to the dried gel pieces and incubate for 30 min at 37°C.
2. Remove the DTT solution and add 100 µl 10 mM iodoacetamide.
3. Incubate for 30 min in the dark at room temperature.
4. Wash the gel pieces twice with 50 mM ammonium bicarbonate pH 8.
5. Dry gel pieces in a centrifugal evaporator prior to proteolytic digestion.

3.2.4. In-Gel Proteolytic Digestion (see Note 9)

1. Rehydrate dried gel pieces in 20 µl of 50 mM ammonium bicarbonate (pH 8) containing 10 ng/µl modified, sequencing grade trypsin (see Note 10).
2. Incubate for approximately 16 h at 37°C.

3.2.5. Preparation for MS/MS Analysis Using a Tandem Mass Spectrometer

1. Spot a 1 µl aliquot of each protease digest on a NP20 ProteinChip Array or on a MALDI target.
2. Add 1 µl of 25% saturated CHCA in 50% ACN, 0.5% TFA.
3. Collect spectra on a tandem MS in reflectron (single MS) mode. While relevant mass ranges will vary depending on

instrumentation, the typical mass range of interest for MALDI TOF-TOF systems is 800–3,500 Da.

4. Compare spectra of the target protein digest to the control digest (see Note 11) to select peaks unique to the protein of interest.

5. Collect MS/MS spectra on selected ions.

6. Submit MS/MS spectra to a database-mining tool such as Mascot (Matrix Sciences) for identification.

3.2.6. Database Searching

The parameters used in database searches depend on the nature of the candidate biomarker and the steps used for purification and enrichment. Critical factors to consider when conducting a database search include:

1. Where known, select the appropriate organism or taxa for the database to query. This accelerates the search and generally provides higher confidence IDs. Search across all taxa in cases where the source is unknown or if proteins foreign to the primary organism are expected (for example, transgenic models, infectious disease, etc.).

2. Select the desired database to search against. The most frequently used databases are NCBI and SwissProt, but other databases may be used as well.

3. Select the appropriate enzyme (trypsin, AspN, etc.).

4. Select global modifications. Global modifications are those that are expected to exist in all cases. The most common global modification encountered in this workflow is reduction and alkylation, which results in a carbamidomethyl modification.

5. Select variable modifications. Common variable modifications include oxidation (M), acetyl (N-term), and propionamide (acrylamide modification of proteins isolated from an SDS-PAGE gel).

6. Select the number of missed cleavages that can be tolerated in the search. This is referred to as "partial" in Mascot and is typically set at 1 or 2.

7. Select mass tolerances for MS (parent ion) and MS/MS spectra. Mass errors of 0.2 Da and 0.7 Da are typically a good starting point for MS and MS/MS, respectively. The range can be widened or narrowed to further optimize the search.

3.3. Identification of Peptides Under ~4 kDa

Reduction and alkylation is optional but is required for positive identification of peptides containing disulfides. Peptides can be tested for disulfides by treating with DTT prior to profiling.

3.3.1. In-Solution Reduction and Alkylation

1. Denature the sample in U9 buffer using 1.5× the sample volume of U9 buffer.

2. Add DTT to a final concentration of 5 mM and incubate for 1 h at room temperature.

3. Add an excess of iodoacetamide and incubate for 15 min at room temperature in the dark.

4. Quench the alkylation reaction with an excess of DTT, incubating for 15 min in the dark.

5. Remove DTT and iodoacetamide using reverse phase chromatography and/or by profiling on ProteinChip arrays. Reverse phase chromatography can be conducted using ZipTips or using the protocol described in Subheading 3.1.3.

3.3.2. Preparation for MS/MS Analysis Using a Tandem Mass Spectrometer

Native peptides can be analyzed from MALDI plates or from chromatrographic or normal phase ProteinChip arrays, depending on sample purity and buffer composition. In some cases, peptides can be enriched directly on chromatographic ProteinChip arrays. In other cases, peptides need to be enriched using one or more of the methods described in Subheading 3.1. For peptides that undergo off-array enrichment steps, reverse phase chromatography is usually used as the final step. These eluates are compatible with MALDI plates and/or NP20 arrays, but can sometimes benefit from an additional on-array enrichment step.

1. Enrich for the peptide of interest by binding to the chromatographic array on which it was originally detected and/or by using column-based methods described in Subheading 3.1.

2. Spot a 1 µl aliquot of each protease digest to the corresponding spot on a ProteinChip array or MALDI target.

3. Add 1 µl of 25% saturated CHCA in 50% ACN, 0.5% TFA.

4. Collect spectra on a tandem MS in reflectron (single MS) mode, focusing on the peptide of interest.

5. Collect MS/MS spectra on the target ion.

6. Submit MS/MS spectra to a database-mining tool such as Mascot (Matrix Sciences) for identification, selecting "no enzyme" as the protease. Adjust other search parameters as described in Subheading 3.2.5.

4. Notes

1. Many plastics release polymers that can alter the MS profile and interfere with protein detection and identification. In particular, avoid siliconized, DEPC-treated, and "low-binding" plastics.

2. Larger sample volumes may be required for more dilute sample types and lower abundance proteins. In the event that the

sample volume exceeds the capacity of the spin column, samples can be incubated in batches in a 15- or 50-ml conical tube and the resin transferred to a spin column during the first wash step.

3. Typically, 1–2 μl of sample is dried directly on an NP20 array and the matrix added, omitting the wash step used for other chemistries. However, high concentrations of salt and detergent (such as those in the anion exchange flow through) will interfere with MS analysis. For these samples, wash spots two to three times with 5–10 μl of water before adding matrix.

4. Reverse phase chromatography is normally performed in batch in a microcentrifuge tube due to significant leakage of most spin columns when using organic solvents.

5. Supernatant should be removed carefully to avoid transferring any reverse phase resin, which can interfere with subsequent processing steps. It is more important to avoid transferring resin than to collect all of the eluate. A 10 μl pipette tip attached to a 1 ml pipette tip can be useful for extracting the supernatant without disturbing the resin.

6. Higher percentage gels should be used for peptides, whereas lower percentage gels should be used for larger proteins. Stains should also be selected with the mass range of interest in mind. In our experience, Instant Blue (Novexin) is particularly useful for staining peptide gels. Silver staining is generally more sensitive that Coomassie-based stains, but can also interfere with downstream analysis.

7. Take care to avoid removing gel pieces during buffer exchange. Gel pieces can be drawn up into the pipette or adhere to the side of the pipette tip during buffer transfer. To help avoid these issues, use a gel loading pipette tip or a 10 μl pipette tip attached to a 1 ml pipette tip to remove the final traces of buffer from the microcentrifuge tube.

8. Extra care is required when drying gel pieces in a centrifugal evaporator to avoid introduction of dust and particulates. Cover tubes with parafilm and make four to six small holes in the film. Also beware of static electricity, which can cause dried gel pieces to fly out of the tube.

9. While in-gel digestion is used most often, proteins from the analytical extraction can also be digested in solution. If performing digestion in solution, take extra care handling gel pieces and buffers to avoid introduction of (human) keratin, which is more common with in-solution digests. For an in-solution digest:

    (a) Remove gel piece or transfer extract to a new microfuge tube.
    (b) Dry protein extract in a centrifugal evaporator.

(c) Re-suspend protein in a 1/40 dilution of 14 M ammonium hydroxide and dry in a centrifugal evaporator.

(d) Resuspend protein in 20 μl of 10 ng/μl of modified, sequencing grade trypsin in 50 mM ammonium bicarbonate (pH 8).

(e) Incubate approximately 4 h at 37°C. If desired, aliquots can be removed at different time intervals and checked to confirm digestion efficiency.

(f) Spot 1 μl of the digest on an NP20 array.

10. While trypsin is the most often used protease for identification, other proteases may also be used. Other enzymes (i.e. AspN, LysC, etc.) are especially useful for confirming the N- and C-termini of modified or truncated proteins. Once the protein has been identified, use a theoretical digest to select a protease that will generate peptides of appropriate lengths for sequence analysis (approximately 10–30 amino acids).

11. Control digests are essential for ensuring the identification of the correct protein target. There are several different types of control digests, depending on the purity of the target protein:

(a) For all gel-based purifications, a blank piece of gel should be extracted and processed alongside the protein target. This is typically extracted from a lane that was not used on the gel to ensure that no protein is present. After processing, the digest of this blank gel piece is used to select peaks that result from autolysis of the protease, which should be excluded from MS/MS analysis and subsequent database searches. Autolysis peaks may be used for internal calibration.

(b) In most cases, samples with high and low levels of the protein target are processed in parallel through all purification steps. These are normally analyzed in parallel on a SDS-PAGE gel to assist in selection and excision of the gel band containing the target protein. This is especially important in this workflow since many proteins do not run true to size on non-reducing gels. Analysis of digests from bands containing high and low levels of the target protein can assist in selecting peaks that are unique to the target protein.

(c) In some cases, the extracted gel bands may contain more than one protein, which makes positive identification of the target protein more challenging. In instances where protein contaminants are observed in the gel extracts, compare the digest of band containing the target protein to that of other gel bands containing the same contaminant in order to select peptides unique to the target protein.

References

1. Peng, J., Stanley, A.J., Cairns, D., Selby, P.J., Banks, R.E.(2009) Using the protein chip interface with quadrupole time-of-flight mass spectrometry to directly identify peaks in SELDI profiles – initial evaluation using low molecular weight serum peaks. *Proteomics* **9**, 492–498.

2. Miguet, L., Bogumil, R., Decloquement, P., Herbrecht, R., Potier, N., Mauvieux, L., Van Dorsselaer, A. (2006) Discovery and identification of potential biomarkers in a prospective study of chronic lymphoid malignancies using SELDI-TOF-MS. *J Proteome Res* **5**, 2258–2269.

3. Brown, A.G., Leite, R.S., Engler, A.J., Discher, D.E., Strauss, J.F. (2006) A hemoglobin fragment found in cervicovaginal fluid from women in labor potentiates the action of agents that promote contraction of smooth muscle cells. *Peptides* **27**, 1794–1800.

4. Simonsen, A.H., McGuire, J, Podust, V.N., Davies, H., Minthon, L., Skoog, I., Andreasen, N., Wallin, A., Waldemar, G., Blennow, K. (2008) Identification of a novel panel of cerebrospinal fluid biomarkers for Alzheimer's disease. Neurobiol Aging. 29, 961–968.

5. Ye, B., Skates, S., Mok, S.C., Horick, N.K., Rosenberg, H.F., Vitonis, A., Edwards, D., Sluss, P., Han, W.K., Berkowitz, R.S., Cramer, D.W. (2006) Proteomic-based discovery and characterization of glycosylated eosinophil-derived neurotoxin and COOH-terminal osteopontin fragments for ovarian cancer in urine. *Clin Cancer Res* **12**, 432–441.

6. Cho, WC., Yip, T.T., Yip, C., Yip, V., Thulasiraman, V., Ngan, R.K., Yip, T.T., Lau, W.H., Au, J.S., Law, S.C., Cheng, W.W., Ma, V.W., Lim, C.K. (2004) Identification of serum amyloid A protein as a potentially useful biomarker to monitor relapse of nasopharyngeal cancer by serum proteomic profiling. *Clin Cancer Res* **10**, 43–52.

7. Zhang, Z., Bast, R.C. Jr., Yu, Y., Li, J., Sokoll, L.J., Rai, A.J., Rosenzweig, J.M., Cameron, B., Wang, Y.Y., Meng, X.Y., Berchuck, A., Van Haaften-Day, C., Hacker, N.F., de Bruijn, H.W., van der Zee, A.G., Jacobs, I.J., Fung, E.T., Chan, D.W.(2004) Three biomarkers identified from serum proteomic analysis for the detection of early stage ovarian cancer. *Cancer Res* **64**, 5882–5890.

8. Vermeulen, R., Lan, Q., Zhang, L., Gunn, L., McCarthy, D., Woodbury, R.L., McGuire, M., Podust, V.N., Li, G., Chatterjee, N., Mu, R., Yin, S., Rothman, N., Smith, M.T.(2005) Decreased levels of CXC-chemokines in serum of benzene-exposed workers identified by array-based proteomics. *Proc Natl Acad Sci* **102**, 17041–17046.

9. Huang, J.T., Leweke, F.M., Oxley, D., Wang, L., Harris, N., Koethe, D., Gerth, C.W., Nolden, B.M., Gross, S., Schreiber, D., Reed, B., Bahn, S. (2006) Disease biomarkers in cerebrospinal fluid of patients with first-onset psychosis, *PLoS Med* **3**, 2145–2158.

10. Melle, C., Bogumil, R., Ernst, G., Schimmel, B., Bleul, A., von Eggeling, F. (2006) Detection and identification of heat shock protein 10 as a biomarker in colorectal cancer by protein profiling. *Proteomics* **6**, 2600–2608.

11. Nguyen, M.T., Dent, C.L., Ross, G.F., Harris, N., Manning, P.B., Mitsnefes, M.M., Devarajan, P.(2008) Urinary aprotinin as a predictor of acute kidney injury after cardiac surgery in children receiving aprotinin therapy. *Pediatr Nephrol* **23**, 1317–1326.

12. Currid, C.A., O'Connor, D.P., Chang, B.D., Gebus, C., Harris, N., Dawson, K.A., Dunn, M.J., Pennington, S.R., Roninson, I.B., Gallagher, W.M. (2006) Proteomic analysis of factors released from p21-overexpressing tumour cells. *Proteomics* **6**, 3739–3753.

13. Guerrier, L., Righetti, P.G., Boschetti, E. (2008) Reduction of dynamic protein concentration range of biological extracts for the discovery of low-abundance proteins by means of hexapeptide ligand library. *Nat Protoc* **3**, 883–890.

14. Sihlbom, C., Kanmert, I., Bahr, H., Davidsson, P. (2008) Evaluation of the combination of bead technology with SELDI-TOF-MS and 2-D DIGE for detection of plasma proteins. *J Proteome Res* **7**, 4191–4198.

15. Boschetti, E., Righetti, P.G. (2008) The ProteoMiner in the proteomic arena: a non-depleting tool for discovering low-abundance species. *J Proteomics* **71**, 255–64.

16. Sennels, L., Salek, M., Lomas, L., Boschetti, E., Righetti, P.G., Rappsilber, J. (2007) Proteomic analysis of human blood serum using peptide library beads. *J Proteome Res* **6**, 4055–4062.

Chapter 5

Biomarker Discovery in Serum/Plasma Using Surface Enhanced Laser Desorption Ionization Time of Flight (SELDI–TOF) Mass Spectrometry

Momar Ndao

Abstract

Proteins and peptides that undergo variations in concentration or state as a result of a biological process or disease may be used as biomarkers for the diagnosis or prognosis of diseases and/or for the monitoring of therapy. Serum/plasma is one of the most easily obtained patient specimens and contains thousands of proteins produced and secreted from cells and tissues. While serum/plasma is a valuable specimen for protein biomarker research, especially in the area of infectious diseases, the dynamic range of the proteome presents a technical challenge. Serum/plasma is dominated by high abundance proteins, such as albumin, immunoglobulins, haptogloblulin, which constitute almost 90% of the total serum/plasma protein by weight and make the detection of the low abundance proteins difficult. Therefore, effective fractionation and separation methods are essential to detect potential biomarker proteins present in small quantities for mass spectrometry.

The current tests for blood-borne protozoan diseases are inadequate by monitoring treatment efficacy or for prognosis and also lack sensitivity and specificity. To overcome these limitations, we began a program to develop novel assays for infectious diseases using mass spectrometric data directly as well as "next generation" assays that exploit the richness of the MS data converted to standard platforms. Here we focus on high-throughput fractionation and proteomic analysis using Surface Enhanced Laser Desorption Ionization Time of Flight (SELDI–TOF) mass spectrometry platform. Separation and enrichment is achieved using stepwise anion exchange fractionation prior to analysis on multiple ProteinChip array chemistries. We have used this approach successfully to identify proteins/peptides or protein "profiles" (biomarkers) in subjects chronically infected with blood-borne protozoan parasites (i.e. Chagas disease, babesia, toxoplasma, malaria), fascioliosis, and cysticercosis.

Key words: Diagnosis, SELDI–TOF MS, Mass spectrometry, Proteomics, Blood safety, Parasitic diseases

1. Introduction

Several parasitic infections including Human African trypanosomiasis, malaria, babesia, and American trypanosomiasis (Chagas disease) are major threats to transfusion services because they can persist up to several years in completely asymptomatic form. Even more striking; some of these parasites can be transmitted from mother to child (e.g. Chagas, malaria) for several generations (e.g. Chagas). The majority of chronically infected individuals are asymptomatic and many become blood donors because they are unaware of their disease status. Blood screening for these protozoa in most developed countries is not mandated and not performed. The current tests are also inadequate for monitoring treatment efficacy and other intervention. Unfortunately the sensitivity and specificity of current methods to screen blood banks can be very poor.

To overcome these limitations, we applied SELDI–TOF MS to identify specific biomarkers for blood-borne parasites. Early users of this technology focused primarily on the identification of biomarkers for cancer (1). Other promising applications have included inflammatory (2–4), allergic (5), and degenerative (6) conditions. The microbiological applications of this technology have been mostly limited to the proteomic analysis of microbial products (7) and in vitro infections (8). However, a small number of investigators have applied SELDI to clinical samples to study infectious conditions. Alpha-defensins were identified as important for HIV non-progressors using SELDI (9). More recently, SELDI has been used in studies of HIV cognitive impairment (10), to monitor clinical status in HBV (11) and HCV (12, 13), to diagnose intrauterine infection and bacterial endocarditis (14), and to study sepsis (15). SELDI is also being applied to tuberculosis diagnostics (16) and SELDI-based "serum fingerprints" associated with SARS (17), fasciolosis in sheep (18), as well as cysticercosis in pig (19, 20, 21).

Serum or plasma samples are the ideal samples for detecting blood-borne parasitic infections, but can be challenging matrices for biomarker discovery due to the wide dynamic range of protein concentrations and the presence of high abundance proteins such as albumin and IgG that can mask lower abundance proteins. Clean-up and fractionation of serum samples prior to downstream mass spectrometry analysis is therefore essential. Several fractionation methods that enable the enrichment and detection of low abundance proteins are available, including single or multidimensional chromatographic separation, immunodepletion, and ligand-based enrichment (ProteoMiner beads). The method described below uses anion exchange chromatography to partition high abundance proteins into specific fractions. Fractions are then profiled on multiple (and orthogonal) ProteinChip array chemistries, providing a second dimension of separation and thereby increasing the number of proteins detected. This method requires only 20 µL of serum or plasma and is amenable to automation. The method described below describes fractionation of 96 serum/plasma samples and can be performed using liquid handling robotics or using a multi-channel pipette and vacuum manifold.

2. Materials

2.1. Serum Fractionation

1. U9 Buffer: 9 M urea, 2% CHAPS, 50 mM Tris–HCl pH 9, 20 mL. Store at −20°C (see Note 1).
2. Rehydration Buffer: 50 mM Tris–HCl, pH 9, 250 mL. Store at 4°C.
3. U1 buffer: 1 M urea, 0.22% CHAPS, 50 mM Tris–HCl pH 9 (dilute U9 buffer 1:9 in rehydration buffer).
4. Wash buffer 1: 50 mM Tris–HCl with 0.1% Octyl-β-D-glucopyranoside (OGP) pH 9, 20 mL. Store at 4°C.
5. Wash buffer 2: 50 mM Hepes with 0.1% OGP pH 7, 30 mL. Store at 4°C.
6. Wash buffer 3: 100 mM NaAcetate with 0.1% OGP pH 5, 30 mL. Store at 4°C.
7. Wash buffer 4: 100 mM NaAcetate with 0.1% OGP pH 4, 30 mL. Store at 4°C.
8. Wash buffer 5: 50 mM NaCitrate with 0.1% OGP pH 3, 30 mL. Store at 4°C.
9. Wash buffer 6: 33.3% isopropanol/16.7% acetonitrile/0.1% trifluoracetic acid, 30 mL. Store at 4°C.
10. 96-well filtration plate filled with dehydrated Q Ceramic HyperD® F sorbent, 1 plate (Bio-Rad, Hercules, CA). Store at 4°C (see Note 2).
11. Microplate sealing strips.
12. V-bottom 96-well microplates (used for samples, fraction collection, and waste), labeled accordingly (see Note 3).
13. Adhesive Sealing Film for Microplates, (e.g. E&K Scientific Cat. No. T396100).
14. Buffer reservoirs.
15. Biomek® 2000 Laboratory Automation Workstation with ProteinChip Biomarker Integration Package (Beckman Coulter) or vacuum manifold (see Note 4).
16. Microplate shaker; preferably a DPC Micromix5 Shaker, which can also be integrated with the Biomek 2000 workstation (Promega) (see Note 5).

2.2. ProteinChipArray Binding

The ProteinChipArrays and related buffers should be purchased from Bio-Rad to obtain reproducible results.

1. CM10 Arrays.
2. CM Low Stringency Buffer: 0.1 M sodium acetate, pH 4.0.
3. IMAC30 Arrays.

4. IMAC Binding Buffer: 0.1 M sodium phosphate, 0.5 M NaCl pH 7.0.
5. IMAC Copper Sulfate Charging Solution: 0.1 M.
6. IMAC Neutralization Buffer: 0.1 M sodium acetate buffer pH 4.
7. H50 Arrays.
8. H50 Buffer: 10% acetonitrile, 0.1% TFA.
9. 50% Methanol or Acetonitrile.
10. Cassette Compatible Bioprocessor Reservoir (see Note 6).
11. Cassette Compatible Bioprocessor.
12. SPA solution: add 200 µL of 100% Acetonitrile and 200 µL of 1% TFA into SPA vial, vortex until fully dissolved (15–20 min) then centrifuge 2 min at maximum speed.

3. Methods

3.1. Serum Fractionation

Serum/plasma and the circulatory proteome is a mixture of high and low abundance proteins, which occur in concentrations ranging from milligrams to picograms, or less, per milliliter. Most biomarkers that carry important diagnostic information are thought to reside in the lower abundance region of the concentration range. The fractionation of the serum/plasma reduces interference by high abundance proteins by reducing sample complexity and therefore facilitates the analysis of lower-abundance proteins and protein isoforms by mass spectrometry. Anion exchange fractionation separates proteins based on their isoelectric point.

3.1.1. Preparation of Q HyperD F Resin

1. Tap the filtration plate on the working bench several times to make sure that all of the dry Q HyperD F beads are settled to the bottom of the plate.
2. Take the filtration plate out of the pouch and carefully remove the top seal on the filtration plate.
3. With an 8-channel pipette, add 200 µL of Rehydration Buffer to each well that you are going to use.
4. Mix the filtration plate on the MicroMix 5 (see Note 7) at room temperature (RT).
5. After 60 min of mixing, place the waste collection plate underneath the filtration plate and apply a vacuum to remove the buffer from the filtration plate.
6. Add 200 µL of Rehydration Buffer to each well.
7. Apply vacuum to remove the buffer in the filtration plate.
8. Repeat steps 6 and 7 three times with Rehydration Buffer and an additional three times with U1 solution (see below for

preparation). Empty the waste plate under the filtration plate as necessary between washes.

9. The resin in the filtration plate is now ready to bind sample. The plate should be used directly after rehydration.

3.1.2. Sample Preparation

1. Bring serum samples to ambient temperature. Spin at $20,000 \times g$ for 10 min at 4°C.
2. Aliquot 20 μL of serum sample to each well of a standard v-bottom 96-well microplate.
3. Add 30 μL of U9 to each well.
4. Cover the microplate with adhesive sealing film and mix on the MicroMix (set at 20, 5, 20) for 20 min at 4°C.
5. While samples are mixing, prepare U1 buffer by adding 10 mL of U9 solution buffer to 80 mL of Rehydration Buffer (50 mM Tris–HCl). The volume is enough for 1 complete plate. If you are using part of the plate, adjust accordingly.

3.1.3. Binding Sample with Sorbent

1. Pipette 50 μL of sample from each well of the sample microplate to the corresponding well in the 96-well filtration plate.
2. Add 50 μL of U1 to each well of the sample microplate. Mix five times.
3. Pipette 50 μL from each well of the sample microplate to the corresponding well in the 96-well filtration plate.
4. Cover the filtration plate with adhesive sealing film and mix on MicroMix (set at 20, 7, 30) for 30 min at 4°C.

3.1.4. Fraction Collection

1. Place the 96-well microplate labeled F1 underneath the filtration plate.
2. Apply the vacuum and collect the flow through into the F1 plate.
3. Add 100 μL of Wash Buffer 1 to each well of the filtration plate.
4. Mix for 10 min on MicroMix 5 (set at 20, 7, 10) at room temperature (RT).
5. Apply the vacuum and collect the eluant into the F1 plate.
6. Aliquot the fraction into the desired number of plates, which were previously labeled (see Note 3).
7. Seal and store at −20°C.
8. For fraction 2, add 100 μL of Wash Buffer 2 to each well of the filtration plate.
9. Mix for 10 min on MicroMix 5 (set at 20, 7, 10) at room temperature (RT).
10. Place the 96-well microplate labeled F2 underneath the filtration plate.

11. Apply the vacuum and collect the eluant into the F2 plate.
12. Repeat steps 8–11. Aliquot fraction as in step 6.
13. Seal and store at −20°C.
14. Repeat steps 8–13 for fractions 3–6.
15. Dispose of the sample plate as biohazard waste if human serum sample is used.
16. If only part of the filtration plate was used, put the filtration plate back into the pouch, seal, and store it at 4°C.

3.2. ProteinChip Array Binding

ProteinChip Arrays provide a variety of surface chemistries that allow researchers to optimize protein capture and analysis. The surface chemistries of the arrays include a series of classic chromatographic chemistries and specialized affinity capture surfaces. Classic chromatographic surfaces include normal phase for generic protein binding; hydrophobic surfaces for reversed-phase capture; cation- and anion-exchange surfaces; and immobilized metal affinity capture (IMAC) for metal-binding proteins. Typically, in the discovery phase, multiple fractions and varieties of ProteinChip arrays are used in order to increase the numbers of protein detected. In the validation phase of biomarker studies, the number of conditions can be typically reduced. Note that because the method described here uses anion exchange chromatography for fractionation, the anion exchange array is not used for profiling. If other fractionation methods are used, the anion exchange (Q10) ProteinChip array should be considered for the profiling step.

3.2.1. CM10 (Weak Cation Exchanger)

1. Place the ProteinChip Array Cassette in the Bioprocessor and add 150–250 µL binding solution to each well. Incubate for 5 min at room temperature with vigorous shaking (i.e. 250 rpm, or on a Micromix, setting 20/7). Repeat once.
2. Remove the buffer from the wells. Immediately add 50–150 µL sample to each well. Recommended concentration: 50–2,000 µg/mL total protein, diluted in binding buffer (Mix 10 µL of sample with 90 µL of binding buffer, see Note 8). Incubate with vigorous shaking for 30 min.
3. Remove the samples from the wells and wash each well with 150–250 µL binding buffer for 5 min, with agitation. Repeat two more times.
4. Remove the binding buffer from the wells and add 150–250 µL de-ionized (DI) water to each well, remove immediately. Repeat once.
5. Remove the reservoir from the Bioprocessor base clamp assembly.
6. Air dry the arrays for 15–20 min.

3.2.2. IMAC30 (Immobilized Metal Affinity Capture)

1. Place the ProteinChip Array Cassette in the Bioprocessor and add 50 μL of 0.1 M copper sulfate solution to each well. Incubate for 10 min at room temperature with vigorous shaking (i.e. 250 rpm, or on a Micromix, setting 20/7).
2. Remove the metal solution from the wells. Immediately add 150–250 μL of de-ionized (DI) water to each well. Incubate for 1 min at room temperature with vigorous shaking.
3. Remove the DI water solution from the wells. Immediately add 150–250 μL of 0.1 M sodium acetate buffer pH 4 (neutralization buffer) to each well. Incubate for 5 min at room temperature with vigorous shaking.
4. Remove the buffer solution from the wells. Immediately add 150–250 μL of DI water to each well. Incubate for 1 min at room temperature with vigorous shaking.
5. Remove the DI water solution from the wells. Immediately add 150–250 μL of binding buffer to each well. Incubate for 5 min at room temperature with vigorous shaking. Repeat once.
6. Remove the binding buffer from the wells. Immediately add 50–150 μL of sample to each well. Recommended concentration: 50–2,000 μg/mL total protein, diluted in binding buffer (Mix 10 μL of sample with 90 μL of binding buffer, see Note 8). Incubate with vigorous shaking for 30 min.
7. Remove the samples from the wells and wash each well with 150–250 μL binding buffer for 5 min, with agitation. Repeat two more times.
8. Remove the binding buffer from the wells and add 150–250 μL DI water to each well, remove immediately. Repeat once.
9. Remove the reservoir from the Bioprocessor base clamp assembly.
10. Air dry the arrays for 15–20 min.

3.2.3. H50 (Reversed Phase)

1. Place the ProteinChip Array Cassette in the Bioprocessor and pre-wash the ProteinChip Arrays by adding 50 μL 50% methanol or acetonitrile for 5 min. Repeat once.
2. Remove the pre-wash solution from the wells and add 150–250 μL binding solution to each well. Incubate for 5 min at room temperature with vigorous shaking (i.e. 250 rpm or on a Micromix, setting 20/7). Repeat once.
3. Remove the buffer from the wells. Immediately add 50–150 μL sample to each well. Recommended concentration: 50–2,000 μg/mL total protein, in binding buffer (Mix 10 μL of sample with 90 μL of binding buffer, see Note 8). Incubate with vigorous shaking for 30 min.

4. Remove the samples from the wells and wash each well with 150–250 μL binding buffer for 5 min, with agitation. Repeat two more times.
5. Remove the binding buffer from the wells and add 150–250 μL de-ionized (DI) water to each well, remove immediately.
6. Remove the reservoir from the Bioprocessor base clamp assembly.
7. Air-dry the arrays for 5–10 min.

3.2.4. Addition of Energy Absorbing Molecule

1. Addition of the energy absorbing molecule (EAM) is performed after removal of the reservoir – use the cassette hold down frame provided with the Cassette Compatible Bioprocessor to keep the cassette flat during EAM addition.
2. Apply 1 μL of SPA solution to each spot. Air dry for 5 min and apply another 1 μL of SPA solution. Allow air-drying.
3. Analyze the arrays using the ProteinChip Reader.

3.3. Data Analysis Using PCS4000

3.3.1. Consolidating Data

1. Organize spectra into folders based on fraction, array type, matrix, and laser energy.
2. Label folders according to project and condition; for example, Project name_F1CSH_Analysis (F1 = Fraction 1, C = CM10 array, S = SPA, L/H = Low/High energy).

3.3.2. Data Pre-processing

1. Perform Baseline and Noise subtraction for each condition
   (a) Baseline subtraction controls how the baseline "noise" (resulting from the way the signal is collected electronically and from chemical noise contributed by EAM) of a selected spectrum is calculated and displayed. Smoothing the baseline improves the accuracy of peak areas and intensities.
   (b) The noise subtraction sets the lower and upper mass limits for which noise should be calculated. Typically, the matrix portion of the mass range (below a mass of 2,000 Da) is excluded from noise calculation.
2. Calibration of spectra for each condition (low and high energy) can be one of two types:
   (a) Internal calibration indicates the currently applied calibration equation was generated from peaks in that spectrum.
   (b) External calibration indicates the currently applied calibration equation was generated from peaks in another spectrum. This method is the most frequently used calibration for profiling, where the identity and thus the mass of most protein species are unknown.

3. Normalization is the process of linearly scaling the intensities of a set of spectra to account for spectrum-to-spectrum variations due to minor variations in protein load, or instrument performance.

   (a) The normalization process calculates the total ion current for each spot, averages them across all spectra, and adjusts the intensity scales for all the spectra to the average total ion current.

   (b) Normalized spectra can then be used for downstream analysis.

   (c) Furthermore, the resulting normalization factors can be used to identify poor quality spectra, as they would demonstrate larger normalization factors.

3.3.3. Data Analysis

Analysis of spectral data requires the creation of clusters from the peaks in the spectra. A cluster is a group of peaks of similar mass that are treated as the same substance across multiple spectra and can therefore be analyzed using statistical tools.

Once the peaks are clustered, the peak intensities from each spectrum can then be used to measure relative expression levels for that substance and differences in expression levels across relevant sample groupings.

1. To create clusters, pre-processed spectra of the same condition (fraction, array type, matrix, and collection settings) should be moved to a separate folder.

2. Critical parameters for clustering should be optimized for each sample. The most important settings include:

   (a) *S/N* and valley depth, which impact the sensitivity of peak detection. We have typically used $S/N = 5$ and valley depth = 3.

   (b) Peak threshold percentage, which defines the percentage of spectra in which the peak must be detected in order to be considered a cluster. This value is usually set at a minimum of 10% to avoid creating clusters based on spurious peaks, but can be increased, particularly when clustering across only two sample groupings.

   (c) Cluster mass window, which refers to the size of the mass window within which a peak is considered to be the same protein species. This is typically set at 0.3% for the lower mass range (up to 25 kDa) and 1–2% for the higher mass range (10–200 kDa).

3. Once clustering is complete, a cluster table is displayed. Each row in the cluster table represents one cluster, and its associated

average mass and various statistics are shown, including the number of peaks and the number of estimated peaks.

4. The *p*-value calculation tests the hypothesis that the medians of the peak intensities of the group are equal. The number of groups is determined by the fields selected. Low *p*-values indicate that there may be a significant difference in expression levels for a particular protein.

Representative data using malaria and control serum samples are shown in Figs. 1–3. The candidate marker (6.6 kDa) detected in malaria subjects but not other diseases (control) are shown.

#	Index	p-value	ROC area	M/Z	M/Z aver...	M/Z median	M/Z STD	M/Z % CV	Intensity
1	73	0.00003436	0.98951049	5899.771	5899.379	5899.771	1.737	0.029	C
2	87	0.00003436	0.98951049	6637.106	6641.682	6641.959	4.501	0.068	68
3	86	0.00003436	0.01048951	6938.092	6943.225	6938.092	11.959	0.172	5
4	88	0.00003436	0.01048951	7138.674	7147.096	7138.674	11.734	0.164	1
5	95	0.00003436	0.01048951	8130.374	8131.297	8130.374	5.487	0.067	6
6	96	0.00003436	0.01048951	8313.558	8311.628	8313.558	18.800	0.226	2
7	8	0.000044167	0.98951049	2787.946	2789.045	2787.946	0.643	0.023	2
8	6	0.000056592	0.98951049	2751.365	2751.461	2751.365	0.816	0.030	9
9	32	0.000056592	0.01048951	3473.837	3473.882	3473.837	3.226	0.093	2
10	97	0.000056592	0.036461538	8604.209	8605.181	8604.299	7.048	0.082	4
11	84	0.00007228	0.961538462	6668.734	6668.734	6668.734	0.000	0.000	28
12	51	0.000092023	0.961538462	4413.934	4413.808	4413.934	1.205	0.027	5
13	65	0.000092023	0.036461538	5370.778	5370.741	5370.778	0.225	0.004	C
14	14	0.000116784	0.961538462	2932.428	2933.581	2932.428	1.380	0.047	2
15	142	0.000116784	0.961538462	23448.203	23454.164	23454.268	18.640	0.079	C
16	101	0.000147737	0.066433566	9287.099	9295.983	9299.260	9.345	0.101	4
17	115	0.000147737	0.961538462	12565.213	12565.647	12565.213	10.702	0.085	C
18	141	0.000147737	0.933566434	22957.611	22954.942	22957.611	12.266	0.053	C
19	48	0.000234177	0.933566434	4277.851	4277.891	4277.851	1.037	0.024	9
20	89	0.000234177	0.066433566	7361.396	7361.186	7361.396	5.883	0.077	C
21	42	0.000293428	0.905594406	3968.467	3968.499	3968.467	0.953	0.024	7
22	43	0.000293428	0.933566434	3984.168	3984.161	3984.168	0.248	0.006	2
23	60	0.000293428	0.066433566	5080.144	5080.163	5080.144	3.637	0.072	4
24	82	0.000293428	0.933566434	6582.228	6582.924	6582.228	2.507	0.038	11
25	93	0.000293428	0.933566434	7853.331	7852.467	7853.331	3.278	0.042	C
26	33	0.000366505	0.066433566	3526.717	3527.382	3526.717	3.703	0.105	1
27	66	0.000366505	0.094405594	5439.095	5439.414	5439.095	1.988	0.037	C
28	90	0.000366505	0.933566434	7474.438	7469.143	7466.024	9.390	0.126	C
29	118	0.000456331	0.933566434	13075.607	13076.619	13075.607	4.396	0.034	C
30	50	0.000566378	0.094405594	4359.018	4359.698	4359.018	1.861	0.043	C
31	79	0.000566378	0.905594406	6440.465	6443.479	6443.065	3.582	0.056	27
32	23	0.000700744	0.905594406	3215.494	3217.102	3215.494	3.153	0.098	1
33	7	0.000864255	0.905594406	2763.488	2764.841	2764.779	3.176	0.115	6
34	22	0.000864255	0.122377622	3187.253	3185.190	3187.253	2.886	0.091	3
35	120	0.000864255	0.094405594	13579.553	13581.032	13579.553	8.226	0.061	C
36	131	0.000864255	0.122377622	17130.996	17114.317	17130.996	28.820	0.168	C
37	102	0.001302274	0.122377622	9511.498	9511.436	9511.498	2.798	0.029	C
38	117	0.001302274	0.877622378	12887.407	12886.647	12887.407	6.146	0.048	C
39	119	0.001302274	0.877622378	13360.026	13352.707	13353.026	16.091	0.121	C
40	28	0.001591055	0.877622378	3313.503	3318.087	3320.121	4.654	0.140	5
41	125	0.001591055	0.877622378	14692.731	14687.419	14686.897	9.861	0.067	C
42	52	0.001937788	0.84965035	4432.500	4432.124	4432.500	1.587	0.035	4
43	137	0.001937788	0.877622378	19920.072	19917.479	19920.072	30.521	0.153	C
44	17	0.002352708	0.178321678	3036.622	3036.146	3036.622	1.189	0.039	C
45	47	0.002352708	0.84965035	4177.379	4179.621	4180.381	3.204	0.077	2
46	123	0.002352708	0.84965035	14112.461	14112.724	14112.461	2.154	0.015	C

Fig. 1. Ciphergen Biomarker Wizard™ software is first used to determine protein clustering across multiple spectra. Next *p*-values are calculated when two or more groups are compared to each other, for example, positively infected sera vs. non-infected (negative) sera. By sorting according to *p*-value, clusters are ranked according to statistical significance. In this example, a ~6.6-kDa biomarker is selected with *p*-value less than 0.0001.

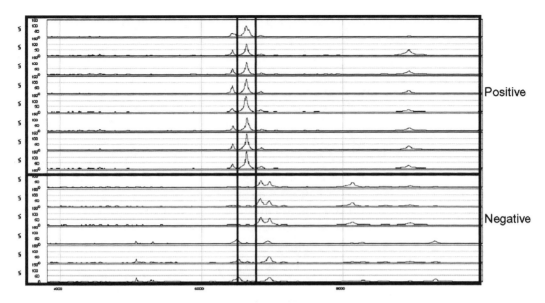

Fig. 2. A "spectra view" is shown of eight positive (*top*) and six negative serum samples (*bottom*) for the biomarker selected in Fig. 1. This particular biomarker found in Fraction 1 (pI > 9) and bound to CM10 arrays was significantly higher in *postive* sera (vs. negative sera: $p < 0.01$).

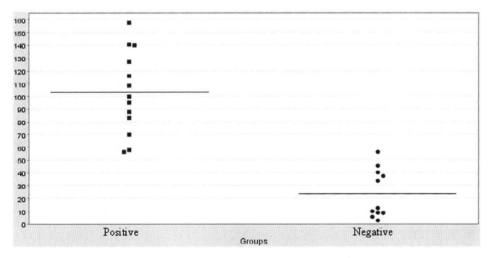

Fig. 3. Ciphergen Express™ v3.0 was used to generate a "cluster plot" for the above biomarker by comparing peak intensities for positive and negative samples. Each *dot* represents a sample and its position on the plot represents the peak intensity of the biomarker in that sample.

4. Notes

1. U9 buffer is sensitive to multiple freeze thaw cycles and should either be made fresh or frozen and thawed only once. For processing small number of samples, aliquot the U9 in 1-mL volumes and store at −20°C.

2. If you use only part of the filtration plate, remove only enough foil seal using a sharp blade to reveal the columns you plan to use. Alternatively, you can remove the whole seal and reseal the wells you are not using with the microplate strips provided. While processing samples, only the wells you are using should be uncovered. After using part of the plate, you should reseal the used wells with the microplate strips provided, and mark those columns as used.

3. The number of plates to be labeled for eluates will vary depending on how many chip types will be used. For example, in our case, we bind three chip types. Therefore we aliquot the fractionated samples into three sets of six v-bottom plates making a total of 18 v-bottom plates to be labeled.

4. A vacuum manifold and manual pipette (or multichannel pipette) may be used as an alternative to liquid handling robotics, although results may not be as reproducible.

5. A wide variety of microplate shakers are available and can be used for this protocol. The MicroMix 5 can be integrated with robotics and reverses direction to ensure efficient mixing. If a microplate shaker is not available, a vortex for microcentrifuge tubes with plate adapter can also be used.

6. In order to avoid cross-contamination do not reuse bioprocessor reservoirs.

7. The settings that yield adequate agitation can vary from one shaker to the next, but should be sufficient to maintain the beads in suspension. We have used settings of form = 20 and amplitude = 7 on our MicroMix 5 shaker.

8. During sample addition, it is important to make sure there are no air bubbles in the wells. To avoid introducing bubbles, lower the pipette tip very close to the spot surface while dispensing sample.

References

1. Henderson N, Steele R. (2005) SELDI–TOF proteomic analysis and cancer detection. *Surgeon.* **3**, 383–90.
2. Gineste C, Ho L, Pompl P, Bianchi M, Pasinetti G. (2003) High-throughput proteomics and protein biomarker discovery in an experimental model of inflammatory hyperalgesia: effects of nimesulide. *Drugs.* **63**, 23–9.
3. Grus F, Joachim S, Pfeifer N. (2003) Analysis of complex autoantibody repertoires by surface-enhanced laser desorption/ionization-time of flight mass spectrometry. *Proteomics.* **3**, 957–61.
4. Yamayoshi Y, Tanabe M, Hoshimo K, MAtsumoto K, Morikawa Y, Shimadzu M et al. (2006) Novel application of ProteinChip technology exploring acute rejection markers of rat small bowel transplantation. *Transplantation.* **82**, 320–6.
5. Hida R, Ohashi Y, Takano Y, Dogru M, Goto E, Fujishima H et al. (2008) Levels of human alpha -defensin in tears of patients with allergic conjunctival disease complicated by corneal lesions: detection by SELDI ProteinChip system and quantification. *Curr Eye Res.* **30**, 723–30.
6. Lewczuk P, Esselmann H, Meyer M, Wollscheid V, Neumann M, Otto M et al. (2003) The amyloid-beta (Abeta) peptide pattern in

cerebrospinal fluid in Alzheimer's disease: evidence of a novel carboxyterminally elongated Abeta peptide. *Rapid Commun Mass Spectrom.* 17, 1291–6.

7. Hess J, Boyle M. (2006) Fibrinogen fragment D is necessary and sufficient to anchor a surface plasminogen-activating complex in Streptococcus pyogenes. *Proteomics.* 6, 375–8.

8. Molestina R, Klein J, Miller R, Pierce W, Ramirez J, Summersgill J. (2002) Proteomic analysis of differentially expressed Chlamydia pneumoniae genes during persistent infection of HEp-2 cells. *Infect Immun.* 70, 2976–81.

9. Zhang L, Yu W, He T, Yu J, Caffrey R, Dalamasso E et al. (2002) Contribution of human alpha-defensin 1, 2, and 3 to the anti-HIV-1 activity of CD8 antiviral factor. *Science.* 298, 1000.

10. Luo X, Carlson K, Wojna V, Mayo R, Biskup T, Stoner J et al. (2003) Macrophage proteomic fingerprinting predicts HIV-1-associated cognitive impairment. *Neurology.* 60, 1931–7.

11. He Q, Lau G, Zhou Y, Yuen S, Lin M, Kung H et al. (2003) Serum biomarkers of hepatitis B virus infected liver inflammation: a proteomic study. *Proteomics.* 3, 666–74.

12. Schwegler E, Cazares L, Steel L, Adam B, Johnson D, Semmes O et al. (2005) SELDI–TOF MS profiling of serum for detection of the progression of chronic hepatitis C to hepatocellular carcinoma. *Hepatology.* 41, 634–42.

13. Gobel T, Vorderwulbecks S, Hauck K, Fey H, Haussinger D, Erhardt A. (2006) New multi protein patterns differentiate liver fibrosis stages and hepatocellular carcinoma in chronic hepatitis C serum samples. *World J Gastroenterol.* 12, 7604–12.

14. Fenollar F, Goncalves A, Esterni B, Azza S, Habib G, Borg JP et al. (2006) A Serum Protein Signature with High Diagnostic Value in Bacterial Endocarditis: Results from a Study Based on Surface-Enhanced Laser Desorption/Ionization Time-of-Flight Mass Spectrometry. *The Journal of Infectious Diseases.* **194**, 1356–66.

15. Nguyen Y, Yaffe M. (2003) Proteomics and systems biology approaches to signal transduction in sepsis. *Crit Care Med.* **31**, S1–6.

16. Bahk Y, Kim S, Euh H, Bai G, Gho S, Kim Y. (2004) Antigens secreted from Mycobacterium tuberculosis: identification by proteomics approach and test for diagnostic marker. *Proteomics.* **4**, 3299–307.

17. Pang R, Poon T, Chan K, Lee N, Chiu R, Tong Y et al. (2006) Serum proteomic fingerprints of adult patients with severe acute respiratory syndrome. *Clin Chem.* 52, 421–9.

18. Rioux MC, Carmona C, Acosta D, Ward B, Ndao M, Gibbs BF et al. (2008) Discovery and validation of serum biomarkers expressed over the first twelve weeks of Fasciola hepatica infection in sheep. *International Journal for Parasitology.* 38, 123–36.

19. Deckers N, Dorny P, Kanobana K, Vercruysse J, Gonzalez AE, Ward B et al. (2008) Use of ProteinChip technology for identifying biomarkers of parasitic diseases: The example of porcine cysticercosis (Taenia solium). *Experimental Parasitology.* 120, 320–9.

20. Ndao M, Spithill TW, Caffrey R, Li H, Podust VN, Perichon R et al. (2010) Identification of Diagnostic Serum Biomarkers for Chagas Disease in Asymptomatic Subjects Using ProteinChip Mass Spectrometry. *Journal of Clinical Microbiology.* 48(4), 1139–49.

21. Ndao M, Rainczuk A, Rioux M, Spithill TW, Ward BJ. (2010) Is SELDI–TOF a valid tool for diagnostic biomarkers? *Trends in Parasitology.* 26(12), 561–7.

Chapter 6

Plasma Proteomic Profiling of Pediatric Osteosarcoma

Yiting Li, Tu Anh Dang, and Tsz-Kwong Man

Abstract

The development of a sensitive, specific, and non-invasive approach for cancer detection will facilitate early detection and, hence, improve the outcome of individuals with known cancer predispositions. Proteomic profiling of blood emerges to be a logical choice of such non-invasive or minimal invasive detection. However, plasma biomarker discovery of pediatric cancers lags behind that of adult cancers, suggesting more efforts are needed in this area. In this study, we used surface-enhanced laser desorption/ionization-time of flight mass spectrometry to profile plasma proteome in osteosarcoma patients. Osteosarcoma is a bone cancer that affects many children and young adults. We have shown that the plasma proteome contains a unique cancer signature that can distinguish patients with osteosarcoma from those with a benign bone disease. To improve cancer biomarker discovery in plasma, we have also shown that depletion of two highly abundant plasma proteins increases the detection sensitivity of lower-abundance proteins. The combination of depletion and proteomic profiling may increase the chance of identifying tumor-derived proteins within the plasma of pediatric cancer patients.

Key words: Proteomics, Osteosarcoma, SELDI, Biomarkers, Plasma, Protein fractionation

1. Introduction

Osteosarcoma is the most common malignant bone tumor in children and young adults, which accounts for approximately 60% of malignant bone tumors in the first two decades of life, with an annual incidence rate of about 5.6 per million (1). It arises from primitive bone-forming mesenchymal cells and is characterized by the production of osteoid material by malignant osteoblastic cells (2). The major cause of death in osteosarcoma is metastasis, which lowers the survival rate of affected patients to only 20%, one-fourth that of patients with localized disease. Furthermore, many patients who present with localized disease are found subsequently to have metastasis, suggesting that micrometastasis had already occurred at diagnosis but was undetected. Therefore, a more sensitive and preferably minimally invasive approach is needed to detect the disease as early as possible, so that the affected children can be treated at a timely basis before the disease spreads to other parts of the body. Epidemiologically, patients with the hereditary form of retinoblastoma (3) and Rothmund-Thomson Syndrome patients with the RECQL4 mutation have much higher risk of developing osteosarcoma (4). Currently, there is no clinical test available that can be used routinely for early detection of osteosarcoma in these patients with known predispositions to the cancer. Hence, an accurate detection method based on plasma proteomic profiles will be useful for monitoring these high-risk patients and early detection of the cancer. To achieve this goal, we tested the feasibility of using surface-enhanced laser desorption/ionization to generate a proteomic signature and classifier in plasma to distinguish osteosarcoma patients from benign osteochondroma patients. The analysis was performed with fractionation of raw plasma, and proteins were captured by weak cationic CM10 chips. In addition, protein mass in plasma is dominated by several highly abundant proteins, such as albumin and immunoglobulins. We also tested if the depletion of these abundant proteins facilitated the detection of lower-abundance proteins in plasma.

2. Materials

2.1. Patients and Samples

Plasma samples from osteosarcoma and osteochondroma patients were analyzed in this study. Plasma samples were obtained through IRB-approved protocols from three collaborative institutions, namely Texas Children's Hospital; Cooks Children's Hospital in Fort Worth, Texas; and Oklahoma Children's Hospital, after informed consent had been signed. All plasma samples were collected in EDTA-containing tubes at room temperature and immediately centrifuged at 1,000 rpm ($100 \times g$) for 10 min. The plasma supernatant was collected and divided into aliquots and stored at −80°C until use.

2.2. Removal of Albumin and IgG from Plasma

ProteoPrep® Blue Albumin and IgG Depletion Kit (Product Code PROTBA, Sigma-Aldrich, St. Louis, MO) is used to deplete albumin and IgG from the plasma samples. Materials included in the kit are:

1. ProteoPrep Blue Albumin and IgG Depletion Medium (Product Code P1120), supplied as a 10-ml suspension containing 60% packed medium.
2. ProteoPrep Blue Equilibration Buffer (Product Code P1245), supplied as a powder. Add 40 ml of high purity water to the

contents of the container to a final volume of 50 ml to make a low ionic strength solution, Tris buffered urea solution, pH 7.8. Aliquot the solution into 1-ml volumes and freeze at −20°C for storage up to 6 months (see Note 1).

3. 25 Spin Columns.
4. 75 Collection Tubes.

2.3. Anion-Exchange Fractionation of Plasma

1. Expression Difference Mapping™ Kit-Serum Fractionation (Ciphergen Biosystems, Fremont, CA) is used for plasma fractionation. Materials included in kit are: 96-well filtration plate filled with dehydrated Q Ceramic HyperD® F sorbent; U9 Buffer (9 M urea, 2% CHAPS, 50 mM Tris–HCl pH 9), 20 ml; Rehydration Buffer (50 mM Tris–HCl, pH 9), 250 ml; Wash buffer 1 (50 mM Tris–HCl with 0.1% OGP pH 9), 20 ml; Wash buffer 2 (50 mM Hepes with 0.1% OGP pH 7), 30 ml; Wash buffer 3 (100 mM NaAcetate with 0.1% OGP pH 5), 30 ml, Wash buffer 4 (100 mM NaAcetate with 0.1% OGP pH 4), 30 ml; Wash buffer 5 (50 mM NaAcetate with 0.1% OGP pH 3), 30 ml; Wash buffer 6 (33.3% isopropanol/16.7% acetonitrile/0.1% trifluoracetic acid), 30 ml; and microplate sealing strips. U9 Buffer should be stored at −20°C while the other regents and buffers are stored at 4°C.
2. V-bottom 96-well microplates.
3. Buffer reservoir.
4. Pipette tips.
5. Biomek® 2000 Laboratory Automation Workstation.

2.4. Plasma Profiling Using CM10 (Weak Cationic Exchanger) ProteinChip® Array

1. CM10 (Weak Cationic Exchanger) ProteinChip® Array (Ciphergen).
2. Cassette Compatible Bioprocessor (Ciphergen).
3. ProteinChip cassette and reservoir (Ciphergen).
4. CM10 Low Stringency binding solution (Ciphergen).
5. SPA (Sinapinic acid) Kit (Ciphergen).
6. Acetonitrile (Sigma).
7. TFA (trifluoroacetic acid) (Sigma).
8. HPLC grade water (Sigma).

2.5. Acquisition of Proteomic Profile

1. Protein Biology System (Model PBSIIc, Ciphergen).
2. ProteinChip® Software 3.2.0.904 (Ciphergen).

2.6. Acquisition of Calibration Spectra

1. NP20 (Normal Phase) Protein Chip Array (Ciphergen).
2. All-in-1 peptide standard, stored at −20°C (Ciphergen).
3. All-in-1 Protein Standard II, stored at −20°C (Ciphergen).

4. CHCA (alpha-cyano-4-hydroxy cinnamic acid) Kit (Ciphergen).
5. SPA (Sinapinic acid) Kit (Ciphergen).
6. Acetonitrile (Sigma).
7. TFA (trifluoroacetic acid) (Sigma).
8. HPLC grade water (Sigma).

2.7. Data Analysis Using Ciphergen Express™ Software

Ciphergen Expression™ Software 3.0.6.001 (Ciphergen).

3. Methods

The sample assignment on a 96-well plate is randomized before the experiment to avoid the spot-to-spot and chip-to-chip variations. The removal of albumin and IgG, as well as fractionation prior to the proteomic experiment improves the peak detection of lower-abundance proteins.

3.1. Removal of Albumin and IgG from Plasma

The protocol is adopted and modified from the Technical Bulletin from ProteoPrep® Blue Albumin and IgG Depletion Kit (Sigma-Aldrich, St. Louis, MO).

1. Gently resuspend the medium by swirling until no settled medium remains on the bottom of the bottle when inverted, and a uniform suspension is formed.

2. Transfer a 0.4-ml aliquot of the suspended medium slurry to a spin column (see Note 2).

3. Centrifuge the spin column in the 2-ml collection tube at 10,000 rpm ($8,000 \times g$) for 5–10 s to remove the storage solution.

4. Add 0.4 ml of Equilibration Buffer to the medium in the spin column and centrifuge at 10,000 rpm ($8,000 \times g$) for 5–10 s. Discard the buffer in the collection tube and place the spin column back into the same collection tube.

5. Add another 0.4 ml of Equilibration Buffer to the medium in the spin column and centrifuge at 10,000 rpm ($8,000 \times g$) for 20–40 s. Discard the buffer in the collection tube and place the spin column into a fresh collection tube.

6. Add the 50 µl of plasma sample to the top of the packed medium bed and incubate at room temperature for 5–10 min. The sample will immediately adsorb into the medium ensuring efficient binding and minimal sample dilution.

7. Centrifuge the spin column and collection tube at 10,000–12,000 rpm ($8,000–12,000 \times g$) for 60 s.

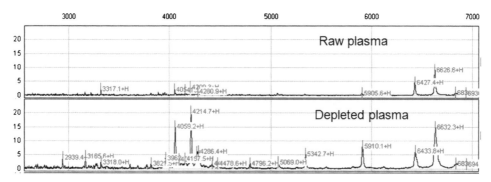

Fig. 1. Enhanced detection of lower-abundance proteins after depletion of albumin and IgG from the plasma. X-axis is the m/z ratio of the protein peak and y-axis is the normalized intensity of the peak. The proteins that were weakly detected or not detected in the raw plasma were readily detected in the albumin and IgG-depleted plasma.

8. Reapply the eluate in the collection tube to the top of the medium bed. Incubate for 5–10 min. This step removes an additional 5% of albumin.

9. Centrifuge the spin column in the same collection tube as before for 60 s. The "two times depleted" serum should remain in the collection tube and should be pooled with the first wash step for optimal protein recovery.

10. Wash the remaining unbound proteins from the spin column by adding 100 µl of Equilibration Buffer to the top of the medium bed, centrifuge for 60 s, and pool with the flow through. The majority (>95%) of the unbound proteins will be in this pool of depleted serum (see Note 3).

11. The albumin/IgG depleted plasma is stored at −80°C for long-term storage. The SELDI-TOF MS spectra of the raw and depleted plasma were shown in Fig. 1.

3.2. Anion-Exchange Fractionation of Plasma

The procedure is adopted and modified from the manual of Expression Difference Mapping™ Kit-Serum Fractionation (Ciphergen).

1. To rehydrate Q HyperD F beads, tap the filtration plate on the working bench several times to ensure that all of the dry Q HyperD F beads have settled onto the bottom of the plate.

2. Take the filtration plate out of the pouch and carefully remove the top seal on the filtration plate.

3. With an eight-channel pipette, add 200 µl of Rehydration buffer to each well of the filtration plate.

4. Mix the filtration plate on the Micro Mix 5 (set at 20, 7, 60) for 60 min at room temperature.

5. Bring plasma samples to ambient temperature and then spin them at $20,000 \times g$ for 10 min at 4°C.

6. For raw plasma, aliquot 25 μl of sample to each well of a standard V-bottom 96-well microplate and add 30 μl of U9 to each well (see Note 4).
7. Cover the microplate with adhesive sealing film and mix on the Micro Mix 5 (Set at 20, 5, 20) at room temperature for 20 min.
8. To prepare reagent U1, add 10 ml of U9 solution to 80 ml of Rehydration Buffer to produce U1 solution.
9. Q HyperD F beads needs to be equilibrated with U1 solution. After step 4, place the waste collection plate underneath the filtration plate and apply vacuum to remove the buffer from the filtration plate. Then, add 200 μl of Rehydration Buffer to each well and apply vacuum to remove the buffer in the filtration plate. Repeat the step three times with Rehydration Buffer and an additional three times with U1 solution. The waste plate under the filtration plate may need to be emptied between washes. The resin in the filtration plate is now ready to bind sample.
10. For raw plasma, pipette 50 μl of sample from each well of the sample microplate to the corresponding well in the 96-well filtration plate and add 50 μl of U1 buffer to each well of the sample plate. Repeat one more time (see Note 5).
11. Cover the filtration plate with adhesive sealing film and mix on Micro Mix 5 (Set at 20, 7, 30) for 60 min at room temperature.

3.3. Fraction Collection

1. For Fraction 1, place the 96-well microplate labeled F1 underneath the filtration plate.
2. Remove the film and apply the vacuum to collect the flow through into the F1 plate.
3. Add 100 μl of Wash Buffer 1 to each well of the filtration plate.
4. Mix for 10 min on Micro Mix 5 (set at 20, 7, 10) at room temperature.
5. Apply the vacuum and collect the eluant into the F1 plate.
6. For Fraction 2, add 100 μl of Wash Buffer 2 to each well of the filtration plate.
7. Mix for 10 min on Micro Mix 5 (set at 20, 7, 10) at room temperature.
8. Place the 96-well microplate labeled F2 underneath the filtration plate.
9. Apply the vacuum and collect the eluant into the F2 plate.
10. Repeat steps 6–9.

11. For Fraction 3, add 100 µl of Wash Buffer 3 to each well of the filtration plate.
12. Mix for 10 min on Micro Mix 5 (set at 20, 7, 10) at room temperature.
13. Place the 96-well microplate labeled F3 underneath the filtration plate.
14. Apply the vacuum and collect the eluant into the F3 plate.
15. Repeat steps 11–14.
16. For Fraction 4, add 100 µl of Wash Buffer 4 to each well of the filtration plate.
17. Mix for 10 min on Micro Mix 5 (set at 20, 7, 10) at room temperature.
18. Place the 96-well microplate labeled F4 underneath the filtration plate.
19. Apply the vacuum and collect the eluant into the F4 plate.
20. Repeat steps 16–19.
21. For Fraction 5, add 100 µl of Wash Buffer 5 to each well of the filtration plate.
22. Mix for 10 min on Micro Mix 5 (set at 20, 7, 10) at room temperature.
23. Place the 96-well microplate labeled F5 underneath the filtration plate.
24. Apply the vacuum and collect the eluant into the F5 plate.
25. Repeat steps 21–24.
26. For Fraction 6, add 100 µl of Wash Buffer 6 to each well of the filtration plate.
27. Mix for 10 min on Micro Mix 5 (set at 20, 7, 10) at room temperature.
28. Place the 96-well microplate labeled F6 underneath the filtration plate.
29. Apply the vacuum and collect the eluant into the F6 plate.
30. Repeat steps 26–29.
31. Seal the six collection microplates and store them at −80°C until use.

3.4. Plasma Profiling Using CM10 (Weak Cationic Exchanger) ProteinChip® Array

1. For fractionated plasma, dilute 20 µl of plasma to 180 µl of binding solution (see Note 6).
2. Place the ProteinChip Array Cassette in the Bioprocessor and add 200 µl binding solution to each well.
3. Incubate for 5 min at room temperature with vigorous shaking (250 rpm). Repeat once.

4. Remove the buffer from the wells and immediately add 200 μl diluted sample to each well. Incubate with vigorous shaking for 60 min.

5. Remove the samples from the wells and wash each well with 200 μl binding solution for 5 min with agitation. Repeat two more times.

6. Remove the binding solution from the wells and add 200 μl of de-ionized water to each well. Remove immediately.

7. Remove cassette and reservoir, aspirate residue water on chip surface by vacuum. Air-dry the array for 5 min.

8. To prepare EAM (Energy Absorbing Molecules, SPA or CHCA), add 5 μl of TFA and 500 μl of acetonitrile to 495 μl of HPLC grade water. Add 400 μl of solvent to each vial of SPA. Vortex and let sit at room temperature for 5 min. Centrifuge at $12,000 \times g$ for 5 min. Aspirate the supernatant to use.

9. Apply 1 μl of EAM (SPA or CHCA) to each spot. Air-dry the spot for 5 min and apply another 1 μl of EAM. Air-dry the spot.

10. Analyze the array using the Protein Biology System.

3.5. Acquisition of Proteomic Profile

1. Start the ProteinChip Software.

2. To create the Spot Protocol of low molecular weight range proteins, first set high mass to 25,000 Da and optimized range from 1,000 Da to 7,500 Da. Next, set the following parameters: starting laser intensity to 170–250, starting detector sensitivity to 4–10 (see Note 7), Focus by optimization center, Mass Deflector to 1,000 Da, and data acquisition method to Seldi Quantitation. For the sample position setting, collect 5 transients per position. Start at position 20 and end at position 80. Move five positions between shots. Totally, 65 transients will be collected. Do not include warming shots. Process sample and identify peaks using auto identify function from 1,000 Da to 10,000 Da.

3. To create the Spot Protocol of medium molecular weight range proteins, set high mass to 200,000 Da, optimized range from 10,000 Da to 50,000 Da. Then set the following parameters: starting laser intensity to 170–250, starting detector sensitivity to 4–10 (see Note 7), Focus mass at 30,000 Da, Mass Deflector to 10,000 Da, and data acquisition method to Seldi Quantitation. For the sample positions setting, collect 5 transients per position. Start at position 22 and end at position 82 (see Note 8). Move five positions between hits. Totally, 65 transients will be collected. Do not include warming shots. Process sample and identify peaks using auto identify function from 10,000 Da to 50,000 Da.

4. To create the Spot Protocol of high molecular weight range proteins, set high mass to 200,000 Da and optimized range

from 10,000 Da to 75,000 Da. Next, set the following parameters: starting laser intensity to 170–250, starting detector sensitivity to 4–10 (see Note 7), Focus by optimization center, Mass Deflector to 10,000 Da, and data acquisition method to Seldi Quantitation. For the sample positions setting, collect 5 transients per position. Start at position 24 and end at position 84 (see Note 8). Move five positions between hits. Totally, 65 transients will be collected. Do not include warming shots. Process sample and identify peaks using auto identify function from 25,000 Da to 75,000 Da.

5. To create Chip Protocol, set Chip Configuration as eight spots lettered A–H and total run as 3. Then, set Spot protocol as "low molecular range" for run 1, "medium molecular range" for run 2, and "high molecular range" for run 3. Next, set first spot as A and last spot as H, increment as every spot, and process order as run. Name the chip protocol and click finish.

6. Click the Sample Exchange Dialog button. Insert the chip when the door opens.

7. Run the Chip Protocol.

8. Save the experiment after finish.

3.6. Acquisition of Calibration Spectra

Two protein standards are used to calibrate the spectra. The All-in-1 peptide is for the calibration of low molecular range proteins and the All-in-1 Protein Standard II is used for the calibration of medium and high molecular weight proteins.

1. To prepare calibration standards, remove the All-in-1 Peptide Standard and All-in-1 Protein Standard II from the freezer and allow them to warm to room temperature. Add 25 μl of HPLC grade water to each vial, flick the tube to mix the liquid, and allow it to sit for 5 min at room temperature. Aliquot the unused solution into 5 μl and store at −20°C.

2. Pipette 5 μl of standard solution onto a spot of an NP20 array. Air-dry the spot completely.

3. Apply 1 μl of CHCA and SPA to the spots of All-in-1 Peptide and All-in-1 Protein II, respectively. Air-dry the spot.

4. Read All-in-1 Peptide Standard.

5. In the software, set high mass to 10,000 Da and optimized range from 1,000 Da to 7,500 Da. Also set the following parameters: starting laser intensity to 170–250, starting detector sensitivity to 4–10 (see Note 7), Focus lag time at 600 ns, Mass Deflector to Auto, and data acquisition method to Seldi Quantitation. For the sample position setting, collect 5 transients per position. Start at position 20 and end at position 80. Move five positions between shots. Set warming positions with five shots at intensity ten units higher than laser intensity. Do not include warming shots.

6. Process sample and identify peaks using auto identify function from 1,000 Da to 7,500 Da.

7. To read All-in-1 Protein Standard II, set high mass to 200,000 Da and optimized range from 10,000 Da to 30,000 Da. Then set the following parameters: starting laser intensity to 170–250, starting detector sensitivity to 4–10 (see Note 7), Focus mass at 15,000 Da, Mass Deflector to 1,000 Da, and data acquisition method to Seldi Quantitation. For the sample position setting, collect 5 transients per position. Start at position 20 and end at position 80. Move five positions between shots. Totally, 65 transients will be collected. Set warming positions with two shots at intensity ten units higher than laser intensity. Do not include warming shots.

8. Process sample and identify peaks using auto identify function from 10,000 Da to 30,000 Da.

3.7. Transfer Data to Ciphergen Express Software

1. Select File\Batch File Export.
2. Select the project or experiments need to be exported.
3. Go to Next and then select Export Spectra to Ciphergen Express System.
4. Click Finish. The files will be copied to Ciphergen Express database.

3.8. Data Processing and Analysis Using Ciphergen Express™ Software

1. Copy spectra collected at the same setting into one folder.
2. Input Sample Name and Sample Group information to each spectrum.
3. Click Subtract Baseline button.
4. To calibrate spectra of low molecular weight, first open spectrum of All-in-1 peptide. Click to label the mass of the peaks with high intensity and Internal Calibration button. Set Calibration Type as 3 parameters. Select weighted in the menu and the calibrants with mass close to the peaks. After that, the user should observe seven peaks in the spectrum, which are: (Arg8)-Vasopressin (1,084.247), Somatostatin (1,637.903), Dynorphin A (209–225) (porcine) (2,147.5), ACTH (1–24) (human) (2,933.5), Insulin B-chain (bovine) (3,495.941), Insulin (human recombinant) (5,807.653), Hirudin BKHV (7,033.614). Then press Calibrate button. Select Spectra\Calibrate\Save Calibration Equation. Name the equation. Highlight spectra to be calibrated. Select Spectra\Calibrate\Select calibration equation and then click Apply.
5. To normalize spectra of low molecular weight, first highlight spectra to be normalized. Click the Normalization button and set the start at beginning at minimum m/z at 1,500. Select End at smallest end and then click Apply.

6. To build cluster of low molecular weight peaks highlight spectra to be clustered. Click the Create EDM button. Set the Peak Detection Parameters as follows: Auto-detect peaks to cluster, first pass as 5.0 of S/N and 5.0 of valley depth, Min Peak Threshold as 0% of all spectra, override threshold for user-detected peaks, Delete user-detected peaks below threshold, and Preserve all first pass peaks. Set Cluster Completion parameters as follows: Cluster Mass Window as 0.7 of Peak width, Second Pass as 2.0 of S/N and 2.0 of valley depth, and Add estimated peaks to complete clusters at cluster center. Specify m/z range from 1,000 to 10,000. Then Press finish and the cluster will build up.

7. To calibrate spectra of medium molecular weight, first open spectrum of All-in-1 Protein Standard II. Click to label of the mass of the peaks with high intensity between 10,000 and 30,000 Da. Click Internal Calibration button. Set Calibration Type as 3 parameters. Select weighted in the menu and the calibrants with mass close to the peaks. The three peaks appeared in the spectrum are: Cytochrome C (bovine) (12,230.92), Myoglobin (equine cardiac) (16,951.51), and Carbonic Anhydrase (bovine RBC) (29,023.662). Press Calibrate button and select Spectra\Calibrate\Save Calibration Equation. Name the equation. Highlight spectra to be calibrated. Select Spectra\Calibrate\Select calibration equation and then press Apply.

8. To normalize spectra of medium molecular weight, highlight spectra to be normalized. Click the Normalization button. Set the start at beginning at minimum m/z at 9,000. Select End at smallest end and then click Apply.

9. To build cluster of medium molecular weight peaks highlight spectra to be clustered. Click the Create EDM button. Set Peak Detection Parameters as follows: Auto-detect peaks to cluster, First pass as 5.0 of S/N and 5.0 of valley depth, Min Peak Threshold as 0% of all spectra, Override threshold for user-detected peaks, Delete user-detected peaks below threshold, and Preserve all first pass peaks. Set Cluster Completion parameters as follows: Cluster Mass Window as 0.7 of Peak width, Second Pass as 2.0 of S/N and 2.0 of valley depth, and Add estimated peaks to complete clusters at cluster center. Specify m/z range from 10,000 to 30,000. Press finish and the cluster will be built up.

10. To calibrate spectra of medium molecular weight, open spectrum of All-in-1 Protein Standard II. Click to label of the mass of the peaks with high intensity between 30,000 and 200,000 Da. Click Internal Calibration button. Set Calibration Type as 3 parameters. Select weighted in the menu and the

calibrants with mass close to the peaks. The three peaks appeared in the spectrum are: Enolase (*S. cerevisiae*) (46,670.92), Albumin (bovine serum) (66,433.0), and IgG (bovine) (147,300.0). Press Calibrate button and select Spectra\Calibrate\Save Calibration Equation. Name the equation. Highlight spectra to be calibrated. Select Spectra\Calibrate\Select calibration equation and then click Apply.

11. To normalize spectra of high molecular weight, highlight spectra to be normalized. Click the Normalization button. Set the start at beginning at minimum m/z at 10,000. Select End at smallest end and then click Apply.

12. To build cluster of high molecular weight peaks highlight spectra to be clustered. Click the Create EDM button. Set Peak Detection Parameters as follows: Auto-detect peaks to cluster, First pass as 5.0 of S/N and 5.0 of valley depth, Min Peak Threshold as 0% of all spectra, Override threshold for user-detected peaks, Delete user-detected peaks below threshold, and Preserve all first pass peaks. Set Cluster Completion parameters as follows: Cluster Mass Window as 0.7 of Peak width, Second Pass as 2.0 of S/N and 2.0 of valley depth, and Add estimated peaks to complete clusters at cluster center. Specify m/z range from 30,000 to 200,000. Press finish and the cluster will be built up.

13. To calculate p-value and Area Under the ROC Curve (AUC), select EDM and then click Calculate P-values. In the window of P-Value Wizard, select the groups to compare. Select paired statistics column and replicates column. Choose the positive group to calculate area under ROC curve and then click finish. The P-value and AUC will be shown in the table for each peak.

14. To export expression values to Excel spreadsheet for analysis using other software (see Note 9), select EDM and then click Export to BPS. In the P-Value Wizard window, select Sample group and replicates column. In the Select Sample Properties to Export window, select spectrum name, sample name, and transpose table, and then click finish. A csv file containing expression values of each peak will be generated. Combine the peak information from different settings of the six fractions into a single spreadsheet.

Using the protein peak information, we identified 19 differentially abundant peaks that are significantly different between the osteosarcoma and osteochondroma samples (Table 1). These 19 peaks constituted a plasma proteomic signature for osteosarcoma. Representative spectra of four differentially abundant peaks were illustrated in Fig. 2. Using these differentially abundant peaks, hierarchical clustering separated the samples into two major groups

Table 1
Protein peaks in plasma that are significantly different between osteosarcoma and osteochondroma patients at the time of diagnosis. (Reproduced from ref. 5 with permission from Wiley-VCH)

				Peak intensity[d]				Fold change
p-value[a]	m/z[b]	Fraction	AUC[c]	OS AVG	OS STD	OC AVG	OC STD	OS/OC
0.0000100	12,691	F1	0.817	1.343	0.759	0.558	0.271	2.407
0.0000119	34,274	F6	0.848	3.904	1.926	2.185	0.591	1.786
0.0000299	42,217	F5	0.167	1.094	0.371	1.748	0.543	0.626
0.0000323	43,229	F5	0.152	1.708	0.64	2.722	0.823	0.627
0.0000746	34,587	F6	0.848	3.678	1.527	2.397	0.504	1.535
0.0000750	42,770	F5	0.167	1.267	0.429	2.021	0.649	0.627
0.0000865	42,433	F5	0.183	1.168	0.394	1.869	0.625	0.625
0.0001003	10,838	F1	0.848	2.577	2.862	0.702	0.261	3.673
0.0001240	21,715	F5	0.167	0.68	0.192	0.951	0.216	0.716
0.0001286	31,925	F6	0.817	2.555	1.908	0.963	0.573	2.654
0.0001374	33,332	F6	0.802	6.492	2.333	4.495	0.989	1.444
0.0002388	108,914	F5	0.194	0.179	0.061	0.253	0.066	0.706
0.0002581	175,716	F2	0.697	0.042	0.011	0.029	0.015	1.444
0.0002915	123,203	F1	0.229	1.099	0.375	1.714	0.599	0.641
0.0003843	11,704	F4	0.833	3.955	8.198	0.295	0.191	13.409
0.0004371	44,468	F5	0.229	1.2	0.323	1.572	0.311	0.763
0.0006519	56,016	F6	0.245	2.553	0.67	3.338	0.771	0.765
0.0007267	10,862	F2	0.771	0.398	0.272	0.21	0.104	1.898
0.0009493	154,713	F1	0.245	0.774	0.337	1.435	0.774	0.539

OS osteosarcoma, OC osteochondroma
[a] p-values were calculated using parametric two-sample t-test
[b] m/z is the mass/charge ratio of the ion peak
[c] AUC (area under the ROC curve) values were calculated using OS as reference
[d] AVG and STD are the average and SD of the average peak intensity of the class

containing osteosarcoma patients or osteochondroma patients (Fig. 3). A proteomic classifier was further developed based on these results and demonstrated that plasma proteomic profiles can be used to distinguish these two groups of patients with high accuracy using leave-one-out cross-validation (5). One of the peaks (m/z 11,700) was identified to be serum amyloid protein A (SAA),

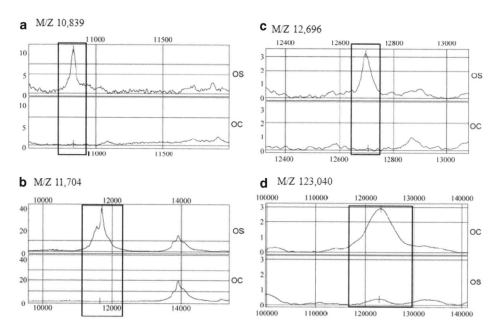

Fig. 2. SELDI-TOF MS spectra of four representative protein peaks that were statistically significant ($p=0.001$) in distinguishing osteosarcoma and osteochondroma. y-Axis is the normalized peak intensity and x-axis is m/z ratio of the peak. (Reproduced from ref. 5 with permission from Wiley-VCH).

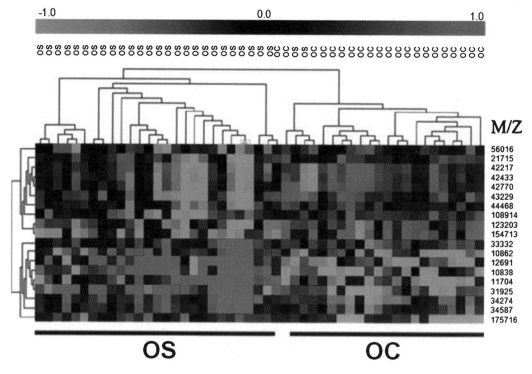

Fig. 3. Hierarchical clustering of osteosarcoma patients using 19 statistically significant protein peaks. Intensities of the peaks were median-centered. Average dot product and average linkage were used in the clustering. *Red to green scale* represents log peak intensity from 21 to 1. Number labels on the *right* are m/z ratios of the 19 significant peaks. *OS* denotes osteosarcoma and *OC* denotes osteochondroma. (Reproduced from ref. 5 with permission from Wiley-VCH).

a common protein in plasma that has been involved in other cancers and diseases. Interestingly, SAA was subsequently found in another study to be involved in the relapse of osteosarcoma patients (6). A recent study has further showed that SAA is expressed in human bone tissues, differentiated osteoblast-like stem cells and human osteosarcoma cell lines (7).

4. Notes

1. The ProteoPrep Blue Equilibration Buffer will become cold after reconstitution and will need to be warmed to 20–25°C to dissolve completely. A 30°C water bath will aid in the dissolution of the material. Do not allow the temperature of the solution to rise above 30°C, since these products may begin to form cyanates, which will be detrimental to proteins.
2. For each spin column, swirl the medium immediately before removing an aliquot to prevent settling of the medium.
3. This fraction may be useful if one wishes to analyze the proteins that bind to albumin and IgG. Some investigators believe that some of these cargo proteins could be used as biomarkers.
4. For the fractionation of depleted plasma, aliquot 80 µl to each well and add 20 µl of Rehydration buffer.
5. For depleted plasma, transfer 100 µl of the depleted sample from each well of the sample plate to the corresponding well in the 96-well filtration plate.
6. For fractionated plasma that has been removed of albumin and IgG, dilute 50 µl of plasma to 150 µl of binding solution.
7. The laser intensity and detector sensitivity should be adjusted to ensure that all the peaks are on the same scale. These two parameters should be optimized for each fraction, since the protein contents are different among the fractions.
8. The laser shot position is shifted between protocols to avoid depletion of signal at certain position.
9. To fully analyze the peak information, we suggest exporting the peaks and their intensities into other more powerful and versatile microarray analysis software. For instance, the data can be imported to TIGR MeV to generate a hierarchical clustering diagram and a heap map of the protein expression in various peaks (Fig. 3) (8). Refer the MeV user manual for the detailed description of the analysis steps (http://www.tm4.org/mev.html).

Acknowledgments

We are indebted to Drs. Ching C. Lau, John Hicks, Murali Chintagumpala at the Texas Children's Cancer Center for their help and support. We would also like to thank Jianhe Shen for her technical support for the study and Carolyn Pena for her assistance in preparing the manuscript. This work was supported by funding from the Carl C. Anderson, Sr. and Marie Jo Anderson Charitable Foundation and the NIH-CA81465.

References

1. Dahlin DC, Unni K. Bone Tumors: General Aspects and Data on 8542 Cases. 4th ed. Springfield, IL: Charles C. Thomas; 1986.
2. Huvos A. Bone Tumors: Diagnosis, Treatment and Prognosis. 2nd ed. Philadelphia: WB Saunders; 1991.
3. Matsunaga E. Hereditary retinoblastoma: host resistance and second primary tumors. J Natl Cancer Inst 1980; 65(1):47–51.
4. Wang LL, Gannavarapu A, Kozinetz CA et al. Association between osteosarcoma and deleterious mutations in the RECQL4 gene in Rothmund-Thomson syndrome. J Natl Cancer Inst 2003; 95(9):669–674.
5. Li Y, Dang TA, Shen J et al. Identification of a plasma proteomic signature to distinguish pediatric osteosarcoma from benign osteochondroma. Proteomics 2006; 6(11): 3426–3435.
6. Jin S, Shen J-N, Guo Q-C et al. 2-D DIGE and MALDI-TOF-MS analysis of the serum proteome in human osteosarcoma. Proteomics Clin Appl 2007; 1:272–285.
7. Kovacevic A, Hammer A, Stadelmeyer E et al. Expression of serum amyloid A transcripts in human bone tissues, differentiated osteoblast-like stem cells and human osteosarcoma cell lines. J Cell Biochem 2008; 103(3):994–1004.
8. Saeed AI, Bhagabati NK, Braisted JC et al. TM4 microarray software suite. Methods Enzymol 2006; 411:134–193.

Chapter 7

Profiling of Urine Using ProteinChip® Technology

Ronald L. Woodbury, Diane L. Bankert McCarthy, and Amanda L. Bulman

Abstract

Urine is an extremely valuable sample type for biomarker discovery due to the non-invasive collection and the relatively low protein content, which makes detection of perturbations associated with disease easier. SELDI-TOF analysis is ideally suited for analysis of urine since the chromatographic capture mechanism can tolerate salt and urea in the urine sample that would otherwise need to be removed prior to mass spectrometric analysis. While neat urine can be analyzed directly on ProteinChip arrays, urine can also benefit from an enrichment step, which has been shown to increase the number of proteins detected more than twofold. Because urine volume and contents can vary substantially between individuals and within individuals over time, sample collection and storage should be carefully controlled to assure reproducible and clinically relevant results.

Key words: Urine, Protein profiling, SELDI, ProteinChip, ProteoMiner enrichment

1. Introduction

Urine has been used for diagnosing disease for centuries, and urinalysis continues to be one of the most commonly used clinical tests. The non-invasive nature of sample collection makes urine an ideal source for diagnostic testing, particularly for indications that may require repeat sampling. Urine is also a relatively simple sample matrix compared to serum, containing less than 150 mg protein/day in healthy individuals. Normal urine volumes range from 0.6 to 2.0 L per day, with most people averaging between 1 and 1.5 L. Increases in total urinary protein, known as proteinuria, are indicative of disease and may result from a glomerular disease, low readsorption rates in the proximal tubule, or an increase in proteins in serum. Proteinuria has been associated with a wide variety of diseases, including diabetes, pre-eclampsia, and multiple myeloma.

The relatively low protein content of urine makes it an ideal sample for detecting perturbations in protein levels associated with disease. However, the low protein concentration and the high salt and urea concentrations encountered in urine make direct analysis challenging by most proteomics techniques. In contrast, SELDI is well suited for direct analysis of urine. Urine samples can be diluted in appropriate binding buffers and applied directly to chromatographic arrays. Salts and urea that interfere with mass spectrometric detection are subsequently washed off, yielding rich protein profiles. SELDI technology has been widely used to analyze urine for a wide variety of diseases, including diabetes (1, 2), renal allograft rejection (3–6), cancer (7–9), renal disease (10), and lupus (11).

While neat urine can be analyzed directly on ProteinChip arrays without pre-treatment, an up-front enrichment step can dramatically increase the number of proteins detected (12). The ProteoMiner technology is most widely used with serum, where it simultaneously reduces the concentration of high abundance proteins and enriches lower abundance proteins, ultimately increasing the number of proteins detected. This technology utilizes a combinatorial peptide ligand library. With sufficient peptide diversity, there should be a binding partner for most, if not all, proteins in a given sample. To achieve optimal function according to this design, the ProteoMiner resin is loaded far in excess of its protein-binding capacity. Under these competitive binding conditions, high affinity interactions with lower abundance proteins occupy more of the binding space on the resin, and most of the higher abundance proteins flow through. These ideal binding conditions are easily achieved when using voluminous samples with high protein concentrations like human serum. Achieving these conditions is more difficult with more dilute samples such as CSF and urine.

Despite the theoretical limitations of ProteoMiner technology when using small volumes of dilute samples, this method has proven extremely effective in increasing the number of protein species detected in clinically relevant volumes of urine. The method described here utilizes a sequential elution scheme, eluting bound proteins from the beads in three steps. Because the relatively high concentrations of salt and urea in urine prevent some proteins from binding to the peptide ligands, the flow through fraction should also be analyzed to maximize the number of proteins detected. Under these conditions, the observed increase in proteins after ProteoMiner treatment is likely a combination of enrichment and sample concentration, but has been shown to yield two to threefold more protein peaks than analysis of neat urine alone.

While urine is a valuable medium for biomarker discovery, urine contents and volume can vary significantly between individuals and within individuals over time. It is therefore especially important to collect samples at the same time of day, using the same methods for collection and pre-storage processing (13).

Several different collection methods are possible, with the most frequently used collections being first and second morning void samples and 24-h urine. Other pre-analytical conditions should be carefully controlled, including the time urine is subjected to room temperature, the number of freeze thaw cycles, and denaturation steps (14). To reduce the impact of variability on statistical analysis, it is also advisable to analyze a larger number of patient samples than might be required for other biological fluids such as serum or CSF.

2. Materials

2.1. Binding of Urine to ProteoMiner Resin

1. 20% (v/v) Methanol in purified water (18 MΩ).
2. ProteoMiner resin, bulk dry (Bio-Rad).
3. Phosphate buffered saline.
4. 2-mL capacity, 96-well plates (Whatman), with silicone plate sealers.
5. Silent screen 96-well filter plate, loprodyne filter 1.2 µm pore (Nunc), and 96-well plates to collect washes.

2.2. Sequential Elution of Urine Proteins into Three Fractions

1. TUC buffer: 7 M thiourea, 2 M urea, 4% CHAPS. Store aliquots of this buffer frozen (−20°C), after thawing discard the unused portion.
2. 96-well v-bottom collection plates (Nunc).
3. UCCit buffer: 9 M urea, 2% CHAPS, 25 mM citric acid – final pH should be ~3.55. Store aliquots of this buffer frozen (−20°C), after thawing discard the unused portion.
4. Organic solvent: 33% isopropyl alcohol, 16.7% acetonitrile, 0.1% trifluoroacetic acid. It is recommended that this volatile solvent be stored cold (4°C) and prepared fresh on a bi-weekly basis.

2.3. Profiling of Fractionated Urine by Retentate Chromatography and MALDI-TOF MS

1. ProteinChip Arrays and bioprocessor apparatus (Bio-Rad).
2. Low stringency binding buffers corresponding to the arrays (Bio-Rad).
3. Sinapinic acid (Bio-Rad), dissolved in 400 µL 50% acetonitrile, 0.1% trifluoroacetic acid by shaking vigorously. Make fresh.

2.4. Profiling of Neat Urine by Retentate Chromatography and MALDI-TOF MS

1. ProteinChip Arrays and bioprocessor apparatus (Bio-Rad).
2. Low stringency binding buffers corresponding to the arrays (Bio-Rad).
3. U9 buffer: 9 M urea, 2% CHAPS, 50 mM Tris–HCl pH 9.
4. Sinapinic acid (Bio-Rad), dissolved in 400 µL 50% acetonitrile, 0.1% trifluoroacetic acid by shaking vigorously. Make fresh.

3. Methods

Unlike other biological fluids like serum and CSF, urine can fluctuate significantly in terms of protein and salt concentration and can exhibit high variability both within a single patient over time and between individuals. Of the approximately 150 mg/day excreted in urine by healthy individuals, approximately 50% is Tamm-Horsfall mucoprotein (uromodulin) and another 10% is serum albumin. An increase in protein in the urine is itself indicative of pathological condition.

3.1. Sample Collection and Storage

1. Due to the high variability of urine over time and between individuals, it is especially important to collect samples at the same time of day and using the same methods.

   (a) 24-h urine compensates for variations over the course of the day and generally yields the most consistent protein content. Where possible, 24-h urine is preferred due to its consistency.

   (b) First and second morning voids are also frequently used. Both have relatively high protein concentrations. First morning voids generally have more particulate matter, and for this reason second morning voids are preferred when available.

2. Urine samples should be centrifuged prior to freezing at −80°C.

3.2. Sample Preparation and Binding to ProteoMiner Resin

1. Swell ProteoMiner beads overnight in 20% methanol. Wash swelled resin once in 5 volumes of water for 5 min with shaking. Equilibrate the beads in PBS by washing several times. Once equilibrated, suspend the beads in an equal volume of PBS to produce a 50% slurry.

2. Thaw urine samples on ice and centrifuge at $16,000 \times g$ for 5 min at 4°C.

3. Add 40 µL of the 50% ProteoMiner slurry (see Note 1) to each well of a multi-well plate (2 mL per well capacity).

4. Add 1.8 mL of urine sample to each well and seal using a silicone plate-sealer. Incubate the plate for 2 h at 4°C with gentle end-over-end rotation.

5. Centrifuge the sample plate $1,000 \times g$ for 1 min to pellet the resin.

6. Aspirate about 1.6 mL of supernatant from each well and save as the Flow Through. Suspend the ProteoMiner beads in the remaining liquid and transfer to a filter plate.

7. Collect the rest of the Flow Through by placing the filter plate on top of a 96-well collection plate and centrifuge $1,000 \times g$ for 1 min. Vacuum filtration may also be used if a centrifuge with a microplate adaptor is not available (see Note 2). Combine with the previously collected Flow Through.

8. Suspend any remaining ProteoMiner beads in the sample plate in 200 μL PBS and transfer to the filter plate. Shake for 5 min and collect the PBS wash by centrifuging at $1,000 \times g$ for 1 min.

9. Add 200 μL PBS to each well of the filter plate, suspend the beads, and collect the wash by centrifugation. The wash may be discarded. Repeat this step three more times.

3.3. Sequential Elution of Urine Proteins into Fractions

Proteins can be eluted from ProteoMiner beads using a variety of different buffers. The method described below elutes proteins in three fractions, but other elution schemes can also be used and are described further in Note 3. Representative profiles of fractionated urine are shown in Fig. 1.

1. Add 40 μL of TUC buffer to each well of the filter plate and shake at room temperature for 15 min.
2. Collect the eluate by centrifugation ($1,000 \times g$ for 1 min).
3. Repeat steps 1 and 2, collecting the eluate in the same collection plate. This is Fraction 1.

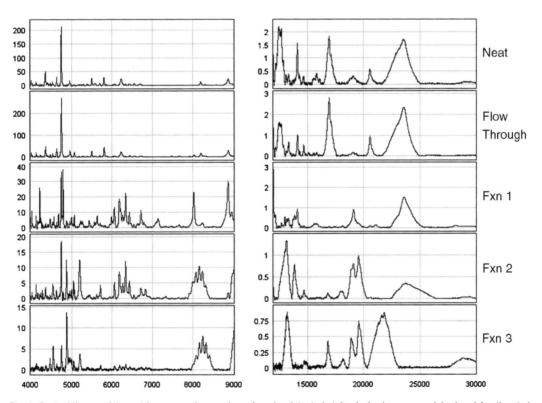

Fig. 1. ProteoMiner enrichment increases the number of peaks detected. 1.8 ml of urine was enriched and fractionated using ProteoMiner beads and the resulting fractions profiled on cation exchange (CM10) arrays. The number of peaks detected in the flow through and three eluates (201 peaks) was approximately 2.5-fold higher than the number detected in neat urine (81 peaks). Data is shown from the peptide range (4–9 kDa) and the protein range (12–25 kDa). Although urine contains relatively few high mass proteins, several proteins are detected only after enrichment.

4. Add 40 μL of UCCit buffer to each well of the filter plate and shake at room temperature for 15 min.

5. Collect the eluate by centrifugation ($1,000 \times g$ for 1 min) in a new collection plate. This is Fraction 2.

6. Repeat steps 4 and 5, collecting the eluate in the same collection plate.

7. Add 40 μL of organic solvent to each well of the filter plate and shake at room temperature for 5 min.

8. Collect the eluate by centrifugation ($1,000 \times g$ for 1 min) in a new collection plate.

9. Repeat steps 7 and 8, collecting the eluate in the same collection plate. This is Fraction 3.

3.4. Profiling of Fractionated Urine by SELDI-TOF MS

1. Assemble the appropriate number of ProteinChip arrays into a bioprocessor. Samples should be profiled in duplicate or triplicate. In addition to experimental samples, a reference sample should be profiled on every array to evaluate the reproducibility of the assay (Fig. 2).

2. Prepare the ProteinChip arrays for profiling. Fractionated urine should be profiled on multiple array chemistries to increase the total number of protein species detected. The most commonly used array chemistries are cation exchange (CM10), anion exchange (Q10), reverse phase (H50), and metal affinity (IMAC30). Representative spectra of neat urine profiles on these four array types are shown in Fig. 3.

    (a) Equilibrate CM10 and Q10 arrays with the appropriate binding buffer by incubating twice for 5 min with 150 μL of the appropriate binding buffer on a plate shaker.

    (b) Pre-treat H50 arrays with 50 μL of 50% acetonitrile for 5 min and allow arrays to dry. Equilibrate by incubating twice for 5 min with 150 μL of H50 binding buffer on a shaker.

    (c) Charge IMAC30 arrays with copper by adding 50 μL of IMAC charging solution and incubating 10 min on a shaker. Remove charging solution and wash once with 150 μL water, incubating 1 min on a shaker. Add 150 μL neutralization buffer, incubating 5 min on a shaker. Remove neutralization solution and wash once with 150 μL water, incubating 1 min on a shaker. Remove water and equilibrate by incubating with IMAC buffer twice for 5 min each on a shaker.

3. Once the chromatographic surface has been equilibrated, add 90 μL binding buffer to each well of the bioprocessor. Add 10 μL of the fraction to be profiled (see Note 4). Shake 30 min at room temperature.

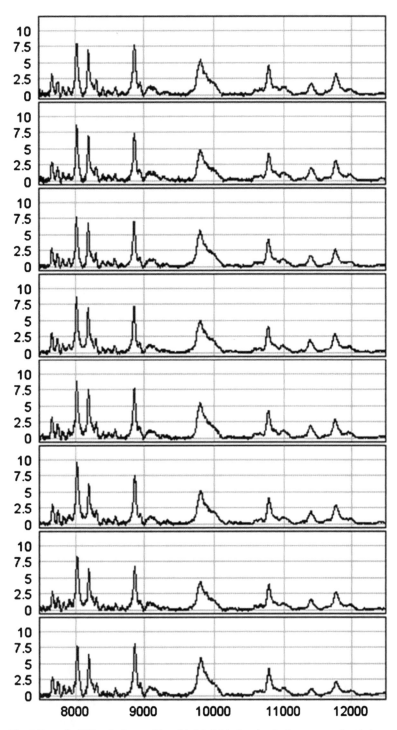

Fig. 2. Reproducibility of urine profiling. A total of 24 aliquots of urine were analyzed alongside experimental samples (one spot on each of 24 ProteinChip arrays) and used to assess assay reproducibility. The data shown here was collected on cation exchange arrays and yielded a median CV of 16.7%. The mass range from 7.5 to 12.5 kDa is shown.

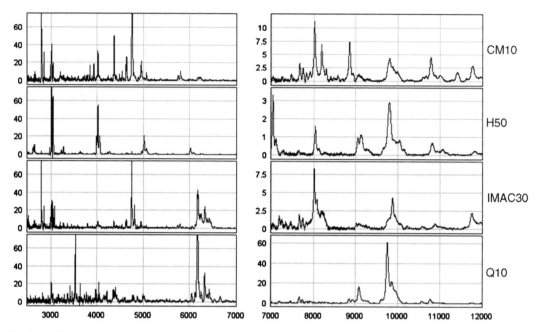

Fig. 3. Profiling on multiple array surfaces increases the number of peaks detected. Neat urine was profiled on four different ProteinChip array chemistries, including cation exchange (CM10), reverse phase (H50), copper affinity (IMAC30-Cu), and anion exchange (Q10). While some proteins bind to multiple surfaces, each surface captures a unique subset of proteins, thereby increasing the total number of peaks detected. Data is shown from two different mass ranges (2.5–7 kDa and 7–12 kDa).

4. Aspirate and discard the liquid from each well. Wash each spot with 150 µL binding buffer for 5 min with vigorous shaking. Repeat this step two more times.

5. Briefly rinse metal affinity and ion exchange arrays with 200 µL of HPLC grade water. Hydrophobic arrays do not require a water wash. Repeat this step. Allow the arrays to dry for 10 min.

6. Apply 1 µL of 12.5 mg/mL SPA (see Note 5) to each spot and allow to dry for 10 min. Repeat this step once.

7. Analyze by TOF MS.

3.5. Profiling of Neat Urine and the ProteoMiner Flow Through Fraction

In the event that sufficient sample volumes are not available to utilize ProteoMiner fractionation, urine samples may be profiled without fractionation. Because urine samples (particularly 24-h samples) are often quite dilute, multiple loadings may be beneficial. As with fractionated urine, samples should be profiled on multiple array types, usually including reverse phase (H50), metal affinity (IMAC30), anion (Q10) and cation (CM10) exchange. Neat urine samples should also be denatured prior to profiling to improve reproducibility and minimize variability between samples.

3.5.1. Sample Preparation

1. Centrifuge urine at $16,000 \times g$ for 5 min at 4°C.
2. Denature urine by mixing 110 µL of urine with 175 µL of U9 buffer.
3. Incubate for 30 min at 4°C.

3.5.2. Profiling

1. Assemble the appropriate number of ProteinChip arrays into a bioprocessor. Samples should be profiled in duplicate or triplicate. In addition to experimental samples, a reference sample should be profiled on every array.
2. Prepare the ProteinChip arrays for profiling. Neat urine should be profiled on multiple array chemistries to increase the total number of protein species detected. The most commonly used array chemistries are reverse phase (H50), metal affinity (IMAC30), and anion (Q10) and cation (CM10) exchange.
   (a) Equilibrate CM10 and Q10 arrays with binding buffer by incubating twice for 5 min with 150 µL of the appropriate binding buffer on a plate shaker.
   (b) Pre-treat H50 arrays with 50 µL of 50% acetonitrile for 5 min and allow arrays to dry. Equilibrate by incubating twice for 5 min with 150 µL of H50 binding buffer on a shaker.
   (c) Charge IMAC30 arrays with copper by adding 50 µL of IMAC charging solution and incubating 10 min on a shaker. Remove charging solution and wash once with 150 µL water, incubating 1 min on a shaker. Add 150 µL neutralization buffer, incubating 5 min on a shaker. Remove neutralization solution and wash once with 150 µL water, incubating 1 min on a shaker. Remove water and equilibrate by incubating with IMAC buffer twice for 5 min each on a shaker.
3. Once the chromatographic surface has been equilibrated, add 175 µL binding buffer to each well of the bioprocessor. Add 25 µL of the neat urine. Shake vigorously for 30 min at room temperature (see Note 6).
4. Aspirate and discard the liquid from each well. Wash each spot with 150 µL binding buffer for 5 min with vigorous shaking. Repeat this step two more times.
5. Briefly rinse metal affinity and ion exchange arrays with 200 µL of HPLC grade water. Hydrophobic arrays do not require a water wash. Repeat this step.
6. Allow the arrays to dry for 10 min.
7. Apply 1 µL of 12.5 mg/mL SPA to each spot and allow to dry for 10 min. Repeat this step once.
8. Analyze by TOF MS.

4. Notes

1. When dispensing aliquots of 50% ProteoMiner bead slurry, vortex the slurry frequently to maintain an even suspension during pipetting.

2. If a centrifuge with a microplate adaptor is not available, a vacuum manifold may also be used for fractionation. Typically, vacuum is applied for 1–2 min at each step. The filtration should be sufficient to remove all liquid from the resin, but overdrying should be avoided since it can make subsequent elution more difficult.

3. A number of different elution schemes can be utilized and selection of the appropriate method depends, in part, on the downstream analysis methods. To elute proteins in a single step, use either UCCit (9 M urea, 2% CHAPS, 25 mM citric acid), or 8 M urea in 5% acetic acid. Two alternative sequential elution schemes are described below. Fractionation using these alternative methods can be performed as described in Subheading 3.

    (a) Scheme 1

Fraction 1: 1 M NaCl, 20 mM HEPES pH 7.5
Fraction 2: 0.2 M Glycine pH 2.4
Fraction 3: 60% ethylene glycol
Fraction 4: Organic solvent (33% isopropyl alcohol, 16.7% acetonitrile, 0.1% trifluoroacetic acid)

    (b) Scheme 2

Fraction 1: 1 M NaCl, 20 mM HEPES pH 7.5
Fraction 2: 3 M Guanidine–HCl, 25 mM MES pH 6
Fraction 3: UCCit buffer (9 M urea, 2% CHAPS, 25 mM citric acid)

4. To improve uniformity of data, it is important to avoid bubbles over the chromatographic spots at the bottom of the wells. Bubbles can be effectively removed by centrifuging the bioprocessor $250 \times g$ for 1 min after addition of buffer.

5. Sinapinic acid is the most frequently used matrix for SELDI-based protein profiling since it allows visualization of the widest mass range. However, other matrices commonly used for protein mass spectrometry may also be used. The second most commonly used matrix for this purpose is α-Cyano-4-hydroxycinnamic acid (CHCA), which is typically made as a 25–50% saturated solution in 0.1% TFA, 50% acetonitrile.

The percentage saturation may need to be adjusted for different mass spectrometers and should be optimized on a reference sample prior to analyzing experimental samples.

6. In some cases, data quality can be improved by applying a second 25 μL of urine to increase protein load. This is especially important for more dilute samples such as 24 h urine.

References

1. Dihazi, H., Muller, G.A., Lindner, S., Meyer, M., Asif, A.R., Oellerich, M., Strutz, F. (2007) Characterization of Diabetic Nephropathy by Urinary Proteomic Analysis: Identification of a Processed Ubiquitin Form as a Differentially Excreted Protein in Diabetic Nephropathy Patients. *Clin Chem.* **53**: *1636–1645.*

2. Out, H.H., Can, H., Spentzos, D., Nelson, R.G., Hanson, R.L., Looker, H.C., Knowler, W.C., Monroy, M., Libermann, T.A., Karumanchi, S.A., Thadhani, R. (2007) Prediction of Diabetic Nephropathy Using Urine Proteomic Profiling 10 Years Prior to Development of Nephropathy. *Diabetes Care.* **30**: *638–643.*

3. O'Riordan, E., Orlova, T.N., Podust, V.N., Chander, P.N., Yanagi, S., Nakazato, M., Hu, R., Butt, K., Delaney, V., Goligorsky, M.S. (2007) Characterization of Urinary Peptide Biomarkers of Acute Rejection in Renal Allografts. *Am J Transplant.* **7**: *930–940.*

4. Schaub,S., Wilkins, J.A., Antonovici, M., Krokhin,O., Weiler,T., Rush, D., Nickerson, P. (2005) Proteomic-Based Identification of Cleaved Urinary beta2-microglobulin as a Potential Marker for Acute Tubular Injury in Renal Allografts. *American Journal of Transplantation* **5**: *729–738.*

5. Schaub, S., Rush, D., Wilkins, J., Gibson, I.W., Weiler, T., (2004) Proteomic-Based Detection of Urine Proteins Associated with Acute Renal Allograft Rejection. *Journal of the American Society of Nephrology.* **15**: *219–227.*

6. Clarke, W., Silverman, B.C., Zhang, Z., Chan, D.W., Klein, A. (2003) Characterization of Renal Allograft Rejection by Urinary Proteomic Analysis. *Annals of Surgery.* **237**: *660–665.*

7. Munro N.P., Cairns, D.A., Clarke, P., Rogers, M., Stanley, A.J., Barrett, J.H., Harnden, P., Thompson, D., Eardley, I., Banks, R.E., Knowles, M.A. (2006) Urinary biomarker profiling in transitional cell carcinoma. *Int J Cancer.* **52**: *2103–2106.*

8. Zhang, Y.F., Wu, D.L., Guan, M., Liu, W.W., Wu, Z., Chen, Y.M., Zhang, W.Z., Lu, Y. (2004) Tree analysis of mass spectral urine profiles discriminates transitional cell carcinoma of the bladder from noncancer patient. *Clin Biochem.* **37**: *772–779.*

9. Vlahou, A., Giannopoulos, A., Gregory, B.W., Manousakas, T., Kondylis, F.I., Wilson, L.L., Schellhammer, P.F., Wright, G.L. Jr., Semmes, O.J. (2004) Protein profiling in urine for the diagnosis of bladder cancer. *Clin Chem.* 2004 **50**: *1438–1441.*

10. Forterre, S., Raila, J., Schweigert, F.J. (2004) Protein profiling of urine from dogs with renal disease using ProteinChip analysis. *J Vet Diagn Invest.* **16**: *271–277.*

11. Mosley, K., Tam, F.W., Edwards, R.J., Crozier, J., Pusey, C.D., Lightstone, L. (2006) Urinary proteomic profiles distinguish between active and inactive lupus nephritis. *Rheumatology* **45**: *1497–1504.*

12. Castagna, A., Cecconi, D., Sennels, L., Rappsilber, J., Guerrier, L., Fortis, F., Egisto Boschetti, E., Lomas, L., Righetti, P.G. (2005) Exploring the Hidden Human Urinary Proteome via Ligand Library Beads. *J Proteome Res.* **4**: *1917–1930.*

13. Papale, M., Pedicillo, M.C., Thatcherm, B.J., Di Paolo, S., Muzio, L.L., Bufo, P., Rocchetti, M.T., Centra, M., Ranieri, E., Gesualdo, L. (2007) Urine profiling by SELDI-TOF/MS: Monitoring of the critical steps in sample collection, handling and analysis. *J Chromatogr B Analyt Technol Biomed Life Sci.* **856**: *205–213.*

Chapter 8

Protein Profiling of Cerebrospinal Fluid

Anja H. Simonsen

Abstract

The cerebrospinal fluid (CSF) perfuses the brain and spinal cord. CSF contains proteins and peptides important for brain physiology and potentially also relevant for brain pathology. Hence, CSF is the perfect source to search for new biomarkers to improve diagnosis of neurological diseases as well as to monitor the performance of disease-modifying drugs. This chapter presents methods for SELDI-TOF profiling of CSF as well as useful advice regarding pre-analytical factors to be considered.

Key words: Cerebrospinal fluid, Proteomic profiling methods, Neurodegeneration, Psychiatry

1. Introduction

The cerebrospinal fluid (CSF) fills the ventricles and surrounds the external surfaces of the brain, bathing and protecting it (1). CSF is mainly produced by the choroid plexus and is directly connected to the extracellular fluid. The extracellular fluid surrounds the neurons and glia, and CSF is therefore believed to reflect brain metabolism (2). In the adult human the total volume of CSF is approximately 140 mL and is produced at a rate of 500 mL per 24 h (3). The protein concentration of CSF is roughly 100 times lower than serum, 300–500 mg/L. It is considered that 80% of this protein content derives from blood (through the choroid tissues) and 20% originates directly from the CNS (4). Changes in CSF composition depend on the blood proteome, CSF circulation alterations, as well as physiological or pathological brain status.

CSF is a rich and relatively accessible source for analyzing changes in brain structural proteins since it is in direct contact with the extracellular space of the brain. Thus changes in brain proteins are likely to be reflected in the CSF. One example of this is the change in the CSF level of total-tau after acute ischemic stroke. CSF-tau is normal for 1–2 days, after which there is a marked increase, which peaks after 2–3 weeks, and then, after 3–4 months there is a return to normal levels (5). The increase most probably reflects leakage of tau from damaged neurons into the CSF, supporting the notion that biochemical analysis of CSF reveals changes that reflect pathogenic processes in the brain. Several studies in CSF have used SELDI successfully for biomarker discovery and validation in neurodegenerative diseases (6–9), psychiatric disorders (10, 11), and brain tumor research (12). We discovered and independently verified biomarkers relevant for different biological processes in the brain such as synaptic proteins, protease inhibitors, and inflammatory proteins that could differentiate between healthy aging and patients with Alzheimer's Disease (AD) (7).

CSF can be collected by lumbar puncture, which has become more routine in recent years. The most common purpose for a lumbar puncture is to collect CSF in a case of suspected meningitis but subarachnoid hemorrhage, hydrocephalus and benign intracranial hypertension may also be supported or excluded by CSF analysis. The incidence of post lumbar puncture headache in patients admitted for diagnostic evaluation of dementia is very low (below 2%), and only occasionally do patients experience more than minimal discomfort after lumbar puncture (13).

CSF concentrations of several proteins are used in clinical practice for the diagnosis of diseases of the brain and nervous system: oligoclonal bands for multiple sclerosis, neuron-specific enolase and 14-3-3 for Creutzfeld Jakob's and S-100b for brain trauma.

The diagnosis of Alzheimer's Disease is aided by analysis of the CSF concentrations of Amyloid beta, Tau and Phosphorylated Tau although the performance of these assays is not optimal, specially with respect to specificity against other dementias (14).

In order to obtain meaningful results from a proteomic analysis of CSF, several aspects of CSF sampling need to be standardized, such as time and volume of CSF collection, sample processing and storage (15).

The method described here does not require pre-treatment of CSF. Anion exchange fractionation was tested during protocol development and, although 15–20% more peaks were visualized many of them were recognized as degradation products from high abundance CSF proteins not seen when profiling neat CSF. Therefore, for most applications, fractionation is not worth the increased sample volume required. CSF protein concentration is approximately 20 times lower than serum, and hence it presents some limitations in sample analysis unless vast volumes are available.

See Fig. 1 for protein profiles across different array surfaces and Fig. 2 for an example biomarker for Alzheimer's disease.

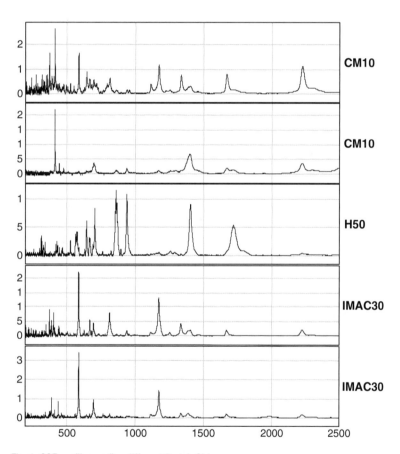

Fig. 1. CSF profiles on five different ProteinChip arrays.

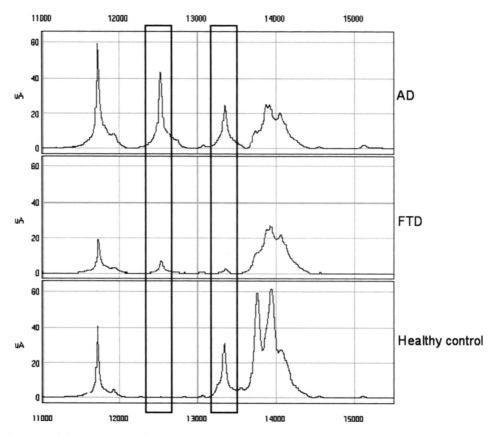

Fig. 2. Representative spectra from the cation exchange array (CM10) showing one Alzheimer's disease (AD) patient, one frontotemporal dementia (FTD) patient, and one healthy control subject. In the *left box*: truncated Cystatin C, in the *right box*: intact Cystatin C.

2. Materials

1. Polypropylene tubes for sample collection and aliquoting, and −80°C resistant labels.
2. ProteinChip arrays: Hydrophobic H50, Metal affinity IMAC30, Cation exchange CM10, Anion exchange Q10.
3. Binding buffers:
   (a) 10% acetonitrile, 0.1% trifluoroacetic acid for H50 arrays.
   (b) 100 mM Copper sulfate, 100 mM Nickel sulfate, 100 mM Sodium acetate (pH 4.0), 100 mM Sodium phosphate (pH 7.0), 0.5 M NaCl for IMAC30-Cu and IMAC30-Ni arrays.
   (c) 100 mM Sodium acetate (pH 4.0) for CM10 arrays.
   (d) 100 mM Tris–HCl (pH 9.0) for Q10 arrays.

4. Energy absorbing matrix: Sinapinic acid (SPA). 50% saturated solution in 50% acetonitrile and 0.1% TFA and alpha-cyano-4-hydroxycinnamic acid (CHCA) 20% saturated solution in 50% acetonitrile and 0.1% TFA.
5. ProteinChip protein calibrant kit.
6. ProteinChip peptide calibrant kit.

3. Methods

3.1. Sample Collection and Storage

CSF samples are usually obtained by lumbar puncture and should be collected at approximately the same time of day (see Notes 1 and 2). Ventricular CSF can also be obtained, usually during neurosurgical procedures or autopsy. CSF should be collected in polypropylene tubes (see Note 3) and centrifuged and aliquoted immediately (see Note 4). CSF samples with macroscopic blood contamination should be discarded immediately after sampling and samples with more than 500 erythrocytes per μL (Fig. 3) should also be discarded (see Note 5). Each CSF sample should be aliquoted into several vials to prevent freeze/thaw cycles and stored at −80°C (see Notes 6 and 7).

Careful consideration should be taken when selecting patients for biomarker discovery studies, age and gender matching are important factors as well as cardiovascular disease that could compromise the blood–brain barrier (see Note 8) and differences in medication use between the investigated groups of patients.

For AD and other dementias, diagnosis is made partly by using tests for decline in the different cognitive domains as well as by brain imaging and neurological examinations. In the very early stages there is considerable overlap between the diseases and between neurodegenerative and psychiatric diseases. Therefore, there is a great need for biomarkers for the early and specific diagnosis of the dementia disorders so that treatment with drugs and other interventions can be started as soon as possible.

By implementing consensus clinical diagnostic criteria in the selection of patients used in our studies (6, 7, 16) we estimated a misdiagnosis rate of 10–20% with some of the elderly controls almost certainly carrying incipient AD pathology. Studies enrolling pathology confirmed AD and Non AD dementia cases and longitudinally followed control individuals are necessary even if they are both costly and time consuming.

3.2. ProteinChip Profiling

In cases where only a limited amount of sample is available, a limited number of analysis conditions can be performed. In that case a prioritized order of array surfaces to use is as follows (most important first): CM10, IMAC30 coupled with Nickel, H50, Q10, and IMAC30 coupled with Copper. SPA should be the first choice of energy-absorbing matrix, with CHCA as the second choice.

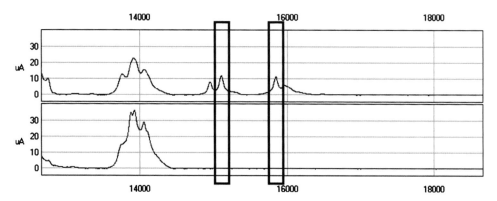

Fig. 3. Spectra from two patients with AD. *Top spectra* shows a CSF sample containing traces of hemoglobin. *In the left box*: Hemoglobin alpha chain. *Right box*: Hemoglobin beta chain. Traces of hemoglobin can sometimes be detected even when there are less than 500 erythrocytes per µl that is the consensus limit for contamination.

Table 1
Typical data acquisition settings for low and high mass range

	Low mass range	High mass range
Mass range	0–100 kDa	0–200 kDa
Focus mass	4 kDa	12 kDa
Detector blanking	1 kDa	2.5 kDa
Number of shots collected	568	568

1. Charge IMAC arrays twice with 50 µL copper or nickel sulfate for 5 min on a shaker at room temperature (RT).
2. Equilibrate ProteinChip arrays twice with 100 µL buffer for 5 min on a shaker at RT.
3. CSF samples should be thawed on ice (see Note 9) and 5 µl of each sample should be diluted into 45 µl of the appropriate binding buffer for each of the ProteinChip® Array types.
4. The samples should be allowed to bind for 60 min at RT.
5. Each array should be washed three times with 100 µL binding buffer.
6. CM10, Q10, and IMAC arrays should be rinsed twice with 100 µL water.
7. Add aliquots of 0.8 µL SPA or CHCA twice.
8. Read arrays on ProteinChip reader at two different instrument settings to focus on lower and higher masses, see Table 1 for typical settings.

To ensure reproducibility of sample preparation and array analysis, a reference CSF standard should be randomly distributed in

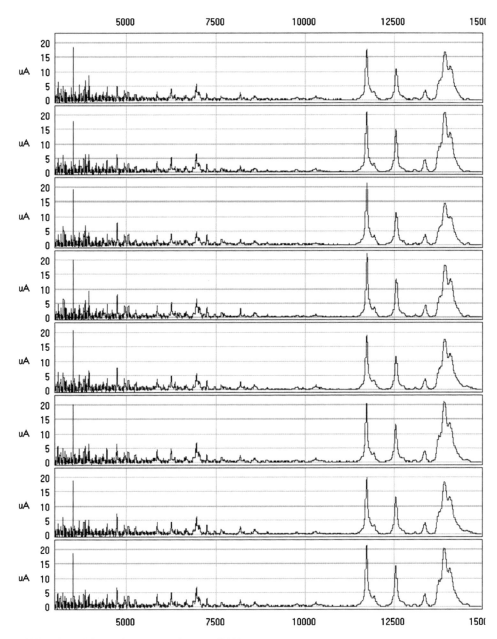

Fig. 4. CSF profiles of eight reference samples on a CM10 array.

several separate aliquots among the clinical samples and analyzed under exactly the same conditions, see Fig. 4 for spectra showing inter array reproducibility. Furthermore, each sample should be run in duplicate or triplicate on separate arrays on successive runs. Reproducibility should be ascertained by measuring the mean CV of peak intensities of automatically detected peaks and is typically between 14 and 19%. Suitable calibration equations should be run using mixtures of calibrants from 3 kDa to 150 kDa for external calibration; internal calibration with known CSF peptides or proteins can also be performed (see Note 10).

4. Notes

Several pre-analytical variables can influence the protein and peptide composition of CSF.

1. CSF sampling should be performed at approximately the same time of day due to circadian fluctuations in some proteins.
2. CSF composition varies with its collection site ventricular or lumbar (17).
3. It is recommended to obtain and store CSF in polypropylene tubes, as some CSF proteins tend to adsorb to other plastic surfaces and glass (18).
4. CSF should be centrifuged ($2,000 \times g$ for 10 min at 4°C) to eliminate cells and other insoluble materials, aliquoted within 4 h after sampling and stored at −80°C (19).
5. Blood contamination can result in misleading profiling of blood proteins; samples should not contain more than 500 erythrocytes per μl.
6. Addition of protease inhibitors is not necessary due to the short time before freezing of the samples and has been shown to decrease the signal in mass spectra (20).
7. Repeated freeze/thaw cycles can also harm the CSF proteins so aliquoting in several portions is advisable.
8. The albumin ratio between CSF and serum should be investigated by standard clinical chemistry as a measure for blood–brain barrier integrity and should be <9. If the blood–brain barrier is damaged there could be increased influx of serum proteins into the CSF.
9. CSF samples should be thawed on ice immediately before application on the ProteinChip arrays. Storage at room temperature for prolonged periods of time after thawing is not recommended, as some proteins will start to degrade.
10. As soon as some of the proteins and peptides in CSF have been purified and identified they can be used for internal calibration before data analysis.

References

1. Segal, M. B. (2000) The choroid plexuses and the barriers between the blood and the cerebrospinal fluid, *Cell Mol. Neurobiol. 20*, 183–196.
2. Segal, M. B. (1993) Extracellular and cerebrospinal fluids, *J. Inherit. Metab Dis. 16*, 617–638.
3. Thompson, E. J. (1995) Cerebrospinal fluid, *J. Neurol. Neurosurg. Psychiatry. 59*, 349–357.
4. Huhmer, A. F., Biringer, R. G., Amato, H., Fonteh, A. N., and Harrington, M. G. (2006) Protein analysis in human cerebrospinal fluid: Physiological aspects, current progress and future challenges, *Dis. Markers. 22*, 3–26.
5. Hesse, C., Rosengren, L., Andreasen, N., Davidsson, P., Vanderstichele, H., Vanmechelen, E., and Blennow, K. (2001) Transient increase in total tau but not phospho-tau in human cerebrospinal fluid after acute stroke, *Neurosci Lett. 297*, 187–190.
6. Simonsen A.H., McGuire J, Podust V.N., Davies H.A., Minthon L., Skoog I., Andreasen N., Wallin A., Waldemar G., and Blennow K (2008) Identification of a Novel Panel of Cerebrospinal Fluid Biomarkers for Alzheimer's Disease, *Neurobiol Aging. 29*, 961–968.
7. Simonsen A.H., McGuire, J., Podust V.N., Hagnelius N-O., Nilsson T.K., Kapaki E., Vassilopoulos D., and Waldemar G. (2007) A novel panel of cerebrospinal fluid biomarkers for the differential diagnosis of Alzheimer's disease versus normal aging and Frontotemporal dementia, *Dement. Geriatr. Cogn Disord. 24*, 434–440.
8. Carrette O., Demalte I., Scherl A., Yalkinoglu O, Corthals G, Burkhard P., Hochstrasser D.F., and Sanchez J.C (2003) A panel of cerebrospinal fluid potential biomarkers for the diagnosis of Alzheimer's disease, *Proteomics. 3*, 1486–1494.
9. Ruetschi, U., Zetterberg, H., Podust, V. N., Gottfries, J., Li, S., Hviid, S. A., McGuire, J., Karlsson, M., Rymo, L., Davies, H., Minthon, L., and Blennow, K. (2005) Identification of CSF biomarkers for frontotemporal dementia using SELDI-TOF, *Exp Neurol. 196*, 273–281.
10. Huang, J. T., Wang, L., Prabakaran, S., Wengenroth, M., Lockstone, H. E., Koethe, D., Gerth, C. W., Gross, S., Schreiber, D., Lilley, K., Wayland, M., Oxley, D., Leweke, F. M., and Bahn, S. (2008) Independent protein-profiling studies show a decrease in apolipoprotein A1 levels in schizophrenia CSF, brain and peripheral tissues, *Mol. Psychiatry. 13*, 1118–1128.
11. Huang, J. T., Leweke, F. M., Tsang, T. M., Koethe, D., Kranaster, L., Gerth, C. W., Gross, S., Schreiber, D., Ruhrmann, S., Schultze-Lutter, F., Klosterkotter, J., Holmes, E., and Bahn, S. (2007) CSF metabolic and proteomic profiles in patients prodromal for psychosis, *PLoS. ONE. 2*, e756.
12. de Bont, J. M., den Boer, M. L., Reddingius, R. E., Jansen, J., Passier, M., van Schaik, R. H., Kros, J. M., Sillevis Smitt, P. A., Luider, T. H., and Pieters, R. (2006) Identification of apolipoprotein A-II in cerebrospinal fluid of pediatric brain tumor patients by protein expression profiling, *Clin. Chem. 52*, 1501–1509.
13. Blennow, K., Wallin, A., and Hager, O. (1993) Low frequency of post-lumbar puncture headache in demented patients, *Acta Neurol Scand. 88*, 221–223.
14. Blennow K. and Hampel H. (2003) CSF markers for incipient Alzheimer's disease, *Lancet Neurol. 2*, 605–613.
15. Giovannoni, G. (2006) Multiple sclerosis cerebrospinal fluid biomarkers, *Dis. Markers. 22*, 187–196.
16. Simonsen, A. H., McGuire, J., Hansson, O., Zetterberg, H., Podust, V. N., Davies, H. A., Waldemar, G., Minthon, L., and Blennow, K. (2007) Novel panel of cerebrospinal fluid biomarkers for the prediction of progression to Alzheimer dementia in patients with mild cognitive impairment, *Arch. Neurol. 64*, 366–370.
17. Roche S, Gabelle A, and Lehmann S (2008) Clinical proteomics of the cerebrospinal fluid: Towards the discovery of new biomarkers, *Proteomics Clin. Appl. 2*, 428–436.
18. Andreasen, N., Hesse, C., Davidsson, P., Minthon, L., Wallin, A., Winblad, B., Vanderstichele, H., Vanmechelen, E., and Blennow, K. (1999) Cerebrospinal fluid beta-amyloid(1–42) in Alzheimer disease: differences between early- and late-onset Alzheimer disease and stability during the course of disease, *Arch. Neurol. 56*, 673–680.
19. Vanderstichele, H., Van, K. E., Hesse, C., Davidsson, P., Buyse, M. A., Andreasen, N., Minthon, L., Wallin, A., Blennow, K., and Vanmechelen, E. (2000) Standardization of measurement of beta-amyloid(1–42) in cerebrospinal fluid and plasma, *Amyloid. 7*, 245–258.
20. Berven, F. S., Kroksveen, A. C., Berle, M., Rajalahti, T., Flikka, K., Arneberg, R., Myhr, K.-M., Vedeler C, Kvalheim, O. M., and Ulvik, R. J. (2007) Pre-analytical influence on the low molecular weight cerebrospinal fluid proteome, *Proteomics Clin. Appl. 1*, 699–711.

Chapter 9

SELDI-TOF Mass Spectrometry-Based Protein Profiling of Tissue Samples for Toxicological Studies

Alexandra Sposny and Philip G. Hewitt

Abstract

Surface-enhanced laser desorption/ionization time-of-flight mass spectrometry (SELDI-TOF MS) has become a popular method for protein profiling in clinical diagnosis, as well as in toxicological studies. It combines solid-phase chromatography with TOF-MS on a single platform, which enables the application of crude samples, such as plasma or tissue lysate. In this chapter, we outline two methods that enable the extraction of proteins from tissue samples for subsequent application on ProteinChip arrays. The first method is the extraction of proteins only from tissue using simply lysis buffers. The second method is helpful if proteins, as well as RNA or DNA, should be extracted from one and the same piece of tissue and is based on phenol–chloroform separation. Although initially developed for liver and kidney tissue both methods can be used for other tissue types.

Key words: Biomarkers, Classification, Hepatotoxicity, Nephrotoxicity, Proteomics, Surface-enhanced laser desorption/ionization time-of-flight mass spectrometry, SELDI

1. Introduction

Molecular toxicology is a discipline that engages genes, proteins and metabolites to study the molecular mode of action of a drug compound as well as to determine its toxicological potential. Protein profiling is one potential method, which examines protein expression differences in biological systems under different conditions. This can be used to improve early prediction of toxicity of a drug candidate and also to elucidate its mode of action. Surface-enhanced laser desorption/ionization time-of-flight mass spectrometry (SELDI-TOF MS) has become a popular method in clinical diagnostics, and to a more limited extent in toxicological studies where it can be used to differentiate between toxic and non-toxic substances. Since SELDI-TOF MS combines chromatography with TOF-MS on one single platform (1, 2), the ProteinChip arrays can be utilized to capture proteins with common biochemical properties while the remaining material can be washed away. The subset of the proteome retained on the array surface can be then directly analyzed by TOF-MS resulting in a "protein profile". This can be described as an accumulation of peaks characterized by their mass/charge (m/z) value on the x-axis and its corresponding intensity on the y-axis (1). These resulting profiles can be used for biomarker identification and subsequent mechanistic investigations. They can also be used for a general protein profiling with subsequent generation of a prediction model based on whole protein profiles.

Besides the most often used plasma samples, molecular methods in toxicology are primarily focused on the analysis of target organ tissue, such as liver or kidney. However, before starting the profiling, proteins have first to be isolated from tissue samples. For liver analysis most often one specified lobe is used. For kidney one has to keep in mind that kidney tissue is not as homogenous as liver tissue is. In kidney we have the cortex (outer part) and the medulla (inner part), which contain different proteins and therefore lead to different protein profiles. Therefore, isolation of proteins should be carried out from half of a kidney to assure the presence of all proteins available in this organ.

It is important to note that the proteins isolated with the protocols described in this chapter are complete protein extracts and do not involve any fractionation or selection from specific organelles.

2. Materials

2.1. Method 1: Complete Protein Extraction from Animal Tissue

2.1.1. Reagents

1. Ice.
2. Liquid nitrogen.
3. Lysis buffer 1: 10 mM Tris–HCl/pH 7.5 (100 μL 1 M), 1 mM EDTA (20 μL 500 mM), 0.2 M Saccharose (685 mg), filled up to final volume of 10 mL ddH$_2$O (see Note 1).
4. Protease Inhibitor Cocktail Set III (Calbiochem).
5. Benzonase (Merck KGaA).
6. Lysis buffer 2: 7 M urea (6.3 g), 2 M thiourea (2.3 g), 4% CHAPS (0.48 g), 40 mM DTT (600 μL 1 M), 20 mM Spermine (300 μL 1 M), filled up to a final volume of 10 mL using ddH$_2$O and stirred at RT until everything is dissolved

completely. Afterwards, final volume will be raised to 12 mL again with ddH$_2$O (see Note 1).

2.1.2. Equipment

1. Ultracentrifuge.
2. Mortar and pestle.
3. Metal spatula.
4. Microfuge tubes (1.5 mL; 2.0 mL).

2.2. Method 2: Extraction of Proteins from Tissue Using TRI Reagent™

2.2.1. Reagents

1. TRI Reagent™ (Sigma-Aldrich).
2. Chloroform, HPLC grade.
3. Isopropanol.
4. 0.3 M Guanidine-hydrochloride/95% Ethanol.
5. Ethanol, HPLC grade.
6. Lysis buffer 2: 7 M urea (6.3 g), 2 M thiourea (2.3 g), 4% CHAPS (0.48 g), 40 mM DTT (600 μL 1 M), 20 mM Spermine (300 μL 1 M), filled up to a final volume of 10 mL using ddH$_2$O and stirred at RT until everything is dissolved completely. Afterwards, final volume will be raised to 12 mL again with ddH$_2$O (see Note 1).
7. Protease Inhibitor Cocktail Set III (Calbiochem).

2.2.2. Equipment

1. Microfuge tubes (1.5 mL; 2.0 mL).
2. Falcon tubes (15 mL).
3. Tissue homogenizer.
4. SpeedVac.

2.3. Sample Application and Mass Calibration

2.3.1. Reagents

1. ProteinChip arrays: CM10, Q10, NP20 (Bio-Rad Laboratories, Hercules, CA, USA).
2. CM10 binding/wash buffer: 100 mM sodium acetate, pH 4 (low stringency).
3. Q10 binding/wash buffer: 100 mM Tris–HCl, pH 9 (low stringency).
4. Acetonitrile (ACN), HPLC grade.
5. Trifluoroacetic acid (TFA), HPLC grade.
6. Sinapinic acid (Bio-Rad Laboratories).

2.3.2. Equipment

1. Shaking plate.
2. Bioprocessor (Bio-Rad Laboratories).
3. ProteinChip reader (Bio-Rad Laboratories).

3. Methods

In the following sections, two methods are described for the isolation of proteins from target organ tissue. The first method is only for protein isolation itself and is carried out using lysis buffers. The second method, which is based on a phenol–chloroform separation, has the advantage that RNA and DNA can in parallel be extracted from the same piece of tissue. This makes the subsequent comparison and correlation of different omics data (genomics, proteomics) easier.

3.1. Method 1: Complete Protein Extraction from Animal Tissue

The volumes of buffers in the following protocol are given for the extraction of liver tissue (100 mg ±50 mg). In case of protein extraction from kidney it should be noted that half of the kidney (weight can vary between 150 mg and 450 mg) should be used to make sure that proteins from the cortex as well as the medulla are included.

3.1.1. First Preparation Steps

1. Cool down centrifuge to 10°C.
2. Place 2.0 mL tubes (see Note 4) on ice to pre-cool (one for each tissue sample).
3. Thaw Inhibitor cocktail and Lysis buffer and put Benzonase on ice.

3.1.2. Extraction of Protein (see Note 2)

1. Take a 1.5 mL tube (see Notes 3 and 4) (pre-frozen by holding in liquid nitrogen for a second) and weigh 100 mg (±50 mg) of tissue.
2. Pulverize tissue into a fine powder using a mortar and pestle (in liquid nitrogen).
3. Take the pre-cooled 2.0 mL tube and add for each tissue sample the following reagents: 125 μL Lysis buffer 1, 30 μL Benzonase (total Benzonase concentration of 25 U/μL=750 U; see Note 5) and 5 μL Protease Inhibitor Cocktail. Immediately add the disrupted tissue using a metal spatula (see Note 2).
4. Resuspend the tissue by pipetting up and down ~30 times.
5. Add 875 μL Lysis buffer 2 and pipette up and down again ~30 times.
6. Vortex suspension.
7. Keep samples on ice until needed (see Note 6).
8. After preparation of all samples they are gently mixed for 1 h at RT.
9. Centrifuge tissue suspensions for 30 min at 10°C and 74,000 × g (ultracentrifuge).

10. Transfer supernatant (containing total cellular protein) into a fresh tube on ice.

11. Samples should be aliquoted and stored frozen at −80°C (see Note 7).

3.2. Method 2: Extraction of Proteins from Tissue Using the TRI Reagent™

This extraction is carried out as per the manufacturer's protocol (3). The protocol outlined below describes the extraction of protein only. Detailed instructions for extraction of DNA/RNA can be obtained from the manufacturer.

3.2.1. First Preparation Steps (see Note 2)

1. Tissue samples are weighed in pre-frozen 1.5 mL tubes (see Notes 3 and 4) (dipped briefly into liquid nitrogen).

2. Add TRI reagent™ (per 10–100 mg of sample use 1 mL TRI Reagent™) directly to the tissue sample and homogenize.

3. Incubate the homogenized samples for 15 min at RT.

4. Add 0.2 mL Chloroform per 1 mL of TRI Reagent™, shake for 15 s and incubate again at RT for 15 min.

5. Centrifuge the mixture at $12,000 \times g$ for 15 min at 4°C.

6. Centrifugation separates the mixture into three phases:

    (a) Colorless, aqueous phase on top containing RNA.

    (b) The interphase in the middle containing DNA.

    (c) The bottom, red organic phase contains protein.

3.2.2. Protein Isolation

1. After careful removal of the two phases containing RNA and DNA, the organic phase containing the protein should be transferred to a 15 mL Falcon tube and mixed with 1.5 mL cold Isopropanol (per mL TRI Reagent™) and shaken carefully.

2. Incubate for 10 min at RT.

3. Centrifuge at $4,000 \times g$ for 20 min at 4°C.

4. Carefully remove the supernatant and wash pellet with 0.3 M Guanidine hydrochloride/95% Ethanol (2.0 mL per 1 mL of TRI Reagent™).

5. Vortex until pellet is loosened and incubate at RT for 20 min.

6. Centrifuge at $4,000 \times g$ for 20 min at 4°C and remove the supernatant carefully.

7. Repeat washing (steps 4–6).

8. After removal of the supernatant, add 2.0 mL Ethanol (100%) and vortex again.

9. Incubate for 20 min at RT.

10. Centrifuge at $4,000 \times g$ for 10 min at 4°C.

11. Aspirate the Ethanol and transfer pellet into a 2.0 mL tube.

12. Dry pellet in SpeedVac but not to total dryness.

13. Resuspend the pellet by pipetting or homogenizing in 300 μL Lysis buffer 2 and 2.5 μL Protease Inhibitor Cocktail.

14. Transfer the solution into a new 2.0 mL tube and suspend rest of sample pellet in further 100 μL Lysis buffer 2 before transferring into the new tube.

15. Centrifuge protein sample at $4,000 \times g$ for 6 min at 4°C and transfer supernatant in a new tube or aliquot.

16. Aliquots can be stored for a short time at −20°C and longer term at −80°C (see Note 7).

3.3. Sample Application

Prior to profiling, method optimization should be performed to determine the optimal conditions for profiling or biomarker detection. During this phase, different Protein Chip surfaces and binding buffers, as well as several sample concentrations, are assessed to define conditions that yield the highest peak count and allow expression differences to be detected. In our toxicology studies, the most often used surfaces for screening of liver and kidney tissue samples are cation-exchange (CM10) and anion-exchange (Q10) surfaces. Because of their different surface chemistries the arrays bind different proteins and the resulting spectra are therefore complementary to each other. Of course, one has to bear in mind that these spectra only reflect a small subset of the whole proteome contained in the original sample. However, with these subsets it is already possible to differentiate between different sample groups (e.g. different time points or doses of treatment).

Sample application can be performed by either direct application "on spot" or with a bioprocessor. During "on spot" analysis, a small volume of buffer/sample solution (~10 μL) is directly applied onto the spot, making this a preferred method for profiling of small sample sizes if only low volumes are available. The use of a bioprocessor (~200 μL per spot) is highly recommended for large-scale experiments to minimize chip-to-chip variation due to differential handling. Furthermore, the use of a bioprocessor allows up to 12 arrays to be processed simultaneously and therefore makes it suitable for higher throughput profiling.

After extraction from tissue and before the application onto the ProteinChip arrays, samples should be assayed for their total protein via Bradford protein quantification. To assure the same amount of protein is applied to each spot (see Note 8) (it is recommended to use ~30 μg protein per spot; see Note 9), samples can be adjusted during the application of protein sample into the appropriate amount of binding buffer. For statistical reasons it is recommended to apply samples in a randomized manner onto the ProteinChip arrays. A short protocol for array preparation (see Note 10) is given below.

3.3.1. Short Protocol: Array Preparation

1. Equilibrate arrays with 150 µL of appropriate buffer for 5 min with mixing, discard buffer and repeat.
2. Apply appropriate volume of binding buffer to each well.
3. Add appropriate volume of sample (equivalent to 30 µg protein) to make a total volume of 150 µL, mix for 45 min, discard buffer.
4. Apply 150 µL binding buffer, mix 5 min, discard buffer and repeat this step twice.
5. Apply 150 µL ddH$_2$O, mix ~30 s, discard ddH$_2$O and repeat.
6. Disassemble bioprocessor and allow spots to dry for a maximum of 15 min (with gentle heating).
7. Apply 1 µL matrix solution SPA (sinapinic acid), allow to dry and repeat application (see Notes 11–13).

3.4. Data Acquisition

Depending on chip surface and sample type, acquisition parameters can vary, and therefore it is necessary to optimize data acquisition settings (e.g., laser energy) before a new experiment is started. To find the best settings for "Data shots" it is advisable to use one or two optimization spots with a mixture of all tissue samples applied that will be acquired within this experiment (same organ). It is helpful to find a laser energy that can be kept throughout the whole experiment without any loss of intensity. Laser energy adjustment is absolutely necessary since sub-optimal laser settings can significantly affect the quality of the resulting spectra. The laser energy could either be set too low leading to an inadequate extraction of proteins from the array surface and a low peak count. In contrast, if the laser energy is set too high, a saturation of the detector is reached resulting in an unstable baseline and broad off-scale peaks, which makes the detection of peak expression differences difficult. Parameters that should be considered during laser energy adjustment are a high peak count with sharp peaks that have high resolution.

Working with sample types and surfaces mentioned in this chapter, it is normally sufficient to cover a mass range of 2,500/3,000 Da up to 35,000 Da. It is also possible to broaden the mass range up to 100,000 Da if marker proteins with higher mass ranges are expected. However, when using such a broad mass range it is advisable to partition the range into two or even three smaller mass ranges in order to optimize for each mass range. More laser energy is generally required to desorb larger proteins. In addition, the focus mass is also an important setting since a higher resolution can be achieved around this mass. Therefore it is recommended either to choose a focus mass in the middle of the acquired mass range in cases where the mass range is narrow (several mass range protocols are applied) or to set the focus mass in the middle of a mass range where most of the peaks are expected. In the case of liver and kidney profiling on cation- and anion-exchange

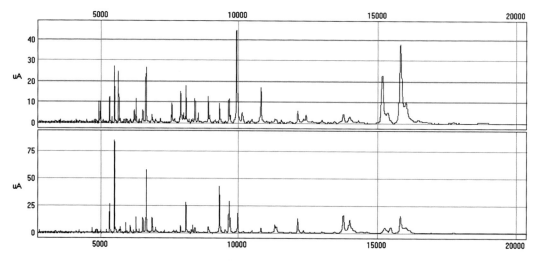

Fig. 1. Liver sample on CM10 (*top*) chip array and Q10 chip array (*bottom*).

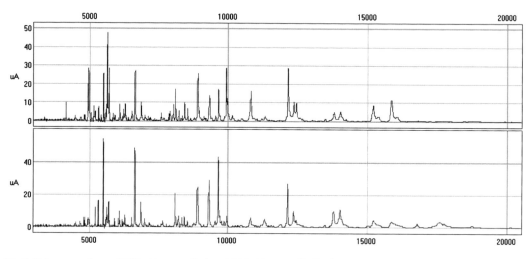

Fig. 2. Kidney samples on CM10 chip array (*top*) and Q10 chip array (*bottom*).

surfaces (as described in this chapter), it is advisable to set the focus mass at around 8,000 Da since most of the acquired peaks are recorded in this range (see Figs. 1 and 2).

Additionally, at least one "Warming shot" should be fired prior the "Data shots", which should have a laser energy ~10% higher than the Data shot laser energy. These additional shots lead to better peak intensities and subsequently improved data quality.

3.5. Mass Calibration   One method of mass calibration of spectra is to spike one or more standard proteins (see Note 14) in the protein sample, which can be directly used for mass and intensity calibration. When using this method, it is necessary to determine whether the spiked protein(s) changes the protein profile in either the number of m/z values or the corresponding intensities. A second method is to generate an external calibration equation (see Note 15) that can be applied to the measured spectra during or after data acquisition.

Experimenters can vary the number and mass range that are used for calibration. However, one has to bear in mind that more calibration points yield better calibration and more precise m/z values in the resulting mass spectra.

4. Notes

1. Lysis buffers 1 and 2 should always be freshly prepared or aliquoted and stored in the freezer at −20°C. Thawed buffer should always be thrown away and should not be refrozen.
2. Take care! Samples should not start thawing at any time.
3. It is important to employ a standard operating procedure for collection and handling of tissue to ensure that differences detected between sample groups are a reflection of the phenotypes being examined rather than a consequence of variations introduced during tissue processing.
4. Use consumables with low protein binding plastics such as polypropylene. Ensure the same consumables are utilized throughout the duration of the study.
5. Benzonase is a recombinant endonuclease that degrades all forms of DNA and RNA.
6. Between each sample processing, mortar and pestle have to be washed with hot water followed by Ethanol and liquid nitrogen.
7. Subjecting samples to several freeze–thaw cycles can generate degradation products and oxidative modifications (+16 Da mass shift) and alter the resulting profile.
8. It is critical to ensure that all samples are identical in regards to protein concentration and the final buffer constituents. The protein concentration affects peak number and intensity. The final buffer constituents of the sample (pH, salt concentration, etc.) affect the binding stringency and resulting profiles.
9. For each new experiment it is important to initially determine the optimal protein concentration by performing a series of replicate titrations. The conditions that yield the highest peak count and reproducibility should be selected. In general it is recommended to use protein extracts within a concentration range of 50–2,000 μg/mL. As the capacity of the array varies for each chip type, depending on the surface chemistry, it is

advantageous to optimize the protein concentration for each type of array.

10. During chip handling use non-latex powder-free gloves such as nitrile. Latex contamination can generate a series of peaks complicating spectra analysis.

11. Preparation of sinapinic acid: Add 200 µL acetonitrile and 200 µL 1% TFA, mix for 15 min at RT, centrifuge 1 min at $8,000 \times g$.

12. Ensure that the drying time of the matrix (sinapinic acid, SPA) does not exceed 20 min, extended incubations can decrease peak intensity and negatively influence assay reproducibility.

13. Use a 2 µL calibrated pipette for matrix addition. Ensure the method of matrix application, including drying time, is consistent as this can affect assay reproducibility.

14. Some of the standard proteins that have been proven to be helpful for calibration are given in Table 1.

15. Preparation of Calibration arrays: Protein standards are prepared according to the package instructions and mixed together to reach the final concentrations (Table 1). The final solution should be carefully mixed and stored in aliquots of 5 µL at −80°C. To prepare the array, 15 µL of SPA solution is added to a 5 µL aliquot of the protein standards and mixed by pipetting gently up and down. 1 µL of this mixture is applied onto each spot of an NP20 array and allowed to air dry. This last step is repeated a second time before the array is ready to use.

Table 1
Standard proteins that should be used for the preparation of a mass calibration array

Standard protein	Mass (Da)	Final concentration (pmol/pL)
Insulin	5,733	0.54
Cytochrome C	12,230	1.08
B-Lactoglobulin A	18,363	2.16
Horseradish peroxidase	43,240	2.16
Serum albumin	66,433	6.49
IgG	147,300	5.41

Given are their molecular weights in Dalton (Da) and the final concentration these proteins should have in the prepared calibration solution

References

1. Wiesner, A. (2004) Detection of tumor markers with ProteinChip technology *Curr. Pharm. Biotechnol.* **5**, 45–67.
2. Tang, N., Tornatore, P., and Weinberger, S. R. (2004) Current developments in SELDI affinity technology *Mass Spectrom Rev.* **23**, 34–44.
3. Extraction Protocol from Sigma-Aldrich using the TRI reagent (http://jackvh.cas.psu.edu/Tri-reagent%20Bulletin.pdf).

Chapter 10

Proteomic Analysis of Skeletal Muscle Tissue Using SELDI-TOF MS: Application to Disuse Atrophy

Mark S.F. Clarke

Abstract

Skeletal muscle atrophy in response to disuse/unloading is a complex adaptation that involves many components of the muscle tissue. The underlying mechanisms that initiate and control the loss of muscle tissue during this response, especially contractile proteins located within the myofibers, are as yet unclear. One approach capable of distinguishing protein changes specifically associated with disuse/unloading-induced skeletal muscle atrophy is to compare the proteomic profiles of similar muscles between control, unloaded/atrophied, and unloaded/"atrophy-protected" experimental conditions. By utilizing a subtractive proteomic analysis approach, those proteins specifically modulated during the atrophic response can be identified and discriminated from those associated with disuse in general. We here describe the use of SELDI-TOF MS coupled with micro-scale preparative ion-exchange chromatography to detect proteins potentially specifically associated with the atrophic response in rat skeletal muscle.

Key words: Skeletal muscle, Atrophy, Disuse, Protein profile, SELDI

1. Introduction

Skeletal muscle atrophy associated with disuse is a complex and as yet poorly understood phenomenon at the cellular and molecular level. Disuse atrophy is a response to a reduction in mechanical loading of the musculoskeletal system and is observed in a variety of populations such as the elderly (1), spinal cord injury patients (2), the chronically bed-ridden (3), cancer patients (4) and astronauts (5). Its etiology is complex and involves adaptations to both the neurological as well as the muscular components of the neuromuscular system.

Numerous investigators have reported both morphological and functional changes in human and animal skeletal tissue as a consequence of disuse (6–11). At the functional level, these changes have included reduced overall muscle volume, as well as decrements in muscle strength and endurance (12, 13). At the morphological/histological level, decreases in overall myofiber cross-sectional area, selective atrophy of fast twitch fiber types based on myosin ATPase activity and fiber type shifting (i.e. fast to slow twitch phenotype, appearance of hybrid myofiber types) have all been reported as a consequence of disuse (13–15). At the biochemical and molecular level, decreases in myosin heavy chain protein expression (15), altered enzymatic profiles (16), reduced levels of total and specific mRNA levels (17–19) have all been reported in atrophied muscle.

More recently, the effects of mechanical unloading on specific biochemical pathways involved in the maintenance of normal muscle mass have been investigated. These have included the role of myostatin (a.k.a. GDF-8) in the skeletal atrophy response to disuse (20, 21), the effect of unloading on the activity of calcium-dependent proteolytic enzyme systems, such as calpain (22) and the ubiquitin/proteosome system (23, 24), and the role of nuclear transcription factors (17), such as NF-kappa-B, in controlling skeletal muscle mass. To date, however, a comprehensive proteomic analysis approach to investigating the underlying pathways involved in disuse skeletal muscle atrophy has yet to be employed.

Established protein analysis techniques capable of resolving complex mixtures of proteins, such as two-dimensional polyacrylamide gel electrophoresis (2D-SDS-PAGE) and high performance liquid chromatography (HPLC), have yielded significant information on the number and the identity of proteins present in both human and rodent skeletal muscle tissue (25, 26). In addition, 2D SDS-PAGE has been employed to resolve changes in the protein expression profiles of skeletal muscle after denervation (27) and immobilization (28) in rodents. However, this approach has several limitations due to the intrinsic properties of muscle tissue and its constituent proteins. The first is the requirement to solubilize all the protein constituents of an inherently dense and fibrous tissue without the use of ionic detergents incompatible with the iso-electric focusing step of 2D SDS-PAGE. The second is the relative abundance of the constituent proteins present in the sample, namely the vast majority being contractile proteins that complicates both the isoelectric focusing and molecular weight-resolving steps of the technique. A third related issue is the inability of existing detection methods to resolve small amounts of protein present in the resolving gel. As such, 2D SDS-PAGE although a powerful technique, remains a laborious and technically challenging approach to protein profiling of skeletal muscle in general, and specifically for identifying those proteins involved in the underlying etiology of disuse atrophy.

In the past, we have used surface-enhanced laser desorption/ionization time-of-flight mass spectroscopy (SELDI-TOF MS) to identify differences in the protein expression profiles of human blood cells exposed to ionizing radiation and reactive oxygen species by way of a subtractive proteomic profiling technique (29). This approach has been successfully used by other investigators to identify protein expression differences between normal and pathological states in an attempt to develop specific biomarkers for these pathological conditions (30–32). This technique can be combined with a number of other protein separation techniques, such tissue fractionation and miniaturized ion exchange chromatography compatible with SELDI-TOF MS, to characterize differences in protein expression between different experimental conditions. As such, SELDI-TOF MS presents itself as a means of resolving complex mixtures of proteins from skeletal muscle tissue, detecting differences in the protein profiles expressed in skeletal muscle during disuse atrophy and providing accurate molecular information to aid in the definitive identification of these proteins. In addition, we have recently shown that disuse atrophy in a rat hind-limb suspension model can be prevented by employing a technique known as dynamic foot stimulation (DFS) (9). By generating a protein expression profile for this "rescue condition" in rat skeletal muscle, we can further identify those proteins selectively modulated within the muscle tissue during disuse atrophy.

2. Materials

2.1. Muscle Samples

Skeletal muscle samples were obtained from a rodent hind-limb suspension (HLS) model of disuse atrophy. Samples were excised from the belly of the soleus muscle of control, HLS (i.e. atrophied muscle) and HLS-DFS (i.e. rescue condition) animals as previously described (9) and immediately frozen in liquid nitrogen in preparation for protein analysis. Soleus muscle was chosen for analysis as this muscle has been shown to atrophy up to 30% in the rat HLS model of disuse atrophy.

2.2. Reagents

All standard laboratory reagents were purchased from Sigma-Aldrich and were of molecular biology grade or above.

2.3. SELDI-TOF MS

Microscale anionic exchange filtration plates and SELDI-TOF MS chips/reagents were all obtained from BioRad Laboratories. A MicroMix 5 plate shaker/mixer and a 96-well plate format vacuum manifold were obtained from Diagnostic Products Corporation. The SELDI-TOF MS Protein Discovery System was purchased BioRad Laboratories.

3. Methods

3.1. Preparation of Muscle Protein Lysates

1. Weigh 10 mg of snap frozen soleus muscle tissue (obtained from control, HLS or HLS-DFS animals) using an analytical balance and place each sample into an individual 2 ml polypropylene microfuge tubes stored on ice.

2. Add a total of 1.5 ml of ice-cold extraction buffer containing 9 M urea, 2% CHAPS, and 50 mM Tris–HCl (aq) (pH 9) to each sample. Also add 15 µl aliquot of Halt™ protease inhibitor cocktail (Pierce Biotechnology) to each sample.

3. Using a micro-scale tissue homogenizer (Fisher Scientific), homogenize each sample thoroughly in the extraction buffer by homogenizing for a total of 30 s with intermittent vortex mixing between 5 s pulses with the homogenizer.

4. After homogenization, centrifuge the sample at $10,000 \times g$ for 10 min at 4°C in a refrigerated micro-centrifuge to pellet any insoluble material.

5. Remove the supernatant and place on ice in preparation for micro-scale ion exchange chromatography of the muscle lysate.

3.2. Microscale Anionic Exchange Chromatography

1. Remove a 96-well microplate format ProteinChip Q anionic exchange filtration plate (BioRad Laboratories) from its packaging and tap the plate on the bench to ensure that the anionic exchange resin contained in each well of the 96-well plate is at the bottom of the well.

2. Add 200 µl of rehydration buffer containing 50 mM Tris–HCl (aq) (pH 9) to each well of the plate using an 8 channel pipette.

3. Seal the plate with an adhesive sealing strip and mix the plate on a MicroMix 5 platform for 60 min at room temperature.

4. After rehydration, place the ProteinChip Q anionic exchange plate on the vacuum manifold and remove the rehydration buffer from the anionic filtration plate by applying a vacuum of 15 in. Hg to the plate for 30 s.

5. Remove ProteinChip Q anionic exchange plate from vacuum manifold and add an additional 200 µl of rehydration buffer containing 50 mM Tris–HCl (aq) (pH 9) to each well of the plate.

6. After 60 s, place the plate on the vacuum manifold and remove the rehydration buffer by applying a vacuum of 15 in. Hg to the plate for 30 s.

7. Repeat steps 5 and 6 above a total of three times to ensure adequate rehydration of the resin contained in each well of the ProteinChip Q anionic exchange plate.

8. After rehydration, apply 200 μl of initial exchange buffer containing 1 M urea, 0.2% (w/v) CHAPS and 5 mM Tris–HCl (aq) (pH 9) to each well of the anionic exchange plate.

9. After a period of 60 s, remove the rehydration buffer by applying a vacuum of 15 in. Hg to the plate for 30 s.

10. Repeat steps 8 and 9 above a total of three times to ensure adequate equilibration of the resin contained in each well of the ProteinChip Q anionic exchange plate with the initial exchange buffer of the resin.

11. The anionic exchange filtration plate is now ready for application of muscle lysate samples.

3.3. Microscale Anionic Exchange Fractionation of Muscle Lysate Samples

1. Take 100 μl of each muscle lysate sample and aliquot into one well of a clean, V-bottomed 96-well plate.

2. Add 150 μl of initial exchange buffer containing 1 M urea, 0.2% (w/v) CHAPS and 5 mM Tris–HCl (aq) (pH 9) to each sample.

3. Seal the sample plate with an adhesive sealing strip and mix for 10 min at 4°C on the MicroMix 5 mixing platform.

4. After mixing of the sample with initial exchange buffer, add 100 μl of each diluted sample to one well in the ProteinChip Q anionic exchange filtration plate.

5. Seal the anionic exchange filtration plate with an adhesive sealing strip and mix for 30 min at 4°C on the MicroMix 5 mixing platform.

6. Remove anionic exchange filtration plate and place on vacuum manifold with a clean, empty V-bottomed 96 well plate (labeled Fraction 1) positioned underneath the exchange plate to collect eluent.

7. Apply vacuum to the anionic exchange filtration plate and collect flow through in the clean, empty V-bottomed 96 well plate.

8. Add 100 μl of Wash Buffer 1 containing 50 mM Tris–HCl (aq) and 0.1% (v/v) *n*-Octyl β-D-glucopyranoside (OGP) (pH 9) to each well of the anionic exchange filtration plate.

9. Seal the sample plate with an adhesive sealing strip and mix for 10 min at 4°C on the MicroMix 5 mixing platform.

10. Place anionic exchange filtration plate on vacuum manifold and again collect eluent into the same V-bottomed 96 well plate as the original flow through. The original flow through and the proteins eluted from the resin at pH 9 become Fraction 1 of the ion exchange protocol. Store Fraction 1 plate at 4°C in preparation for SELDI-TOF MS analysis.

11. Remove the anionic exchange filtration plate from the vacuum manifold and add 100 µl of Wash Buffer 2 containing 50 mM HEPES (aq) and 0.1% (v/v) *n*-Octyl β-D-glucopyranoside (OGP) (pH 7) to each sample well.
12. Seal the sample plate with an adhesive sealing strip and mix for 10 min at 4°C on the MicroMix 5 mixing platform.
13. After mixing, place anionic exchange filtration plate and place on vacuum manifold with a clean, empty V-bottomed 96 well plate (labeled Fraction 2) positioned underneath the exchange plate to collect eluent.
14. Apply vacuum to the anionic exchange filtration plate and collect the eluent.
15. Add an additional 100 µl of Wash Buffer 2 to each sample well. Seal the sample plate with an adhesive sealing strip and mix for 10 min at 4°C on the MicroMix 5 mixing platform.
16. Place anionic exchange filtration plate on vacuum manifold and again collect the eluent into the same V-bottomed 96 well plate as the original Fraction 2. Seal the plate with an adhesive strip and store Fraction 2 plate at 4°C in preparation for SELDI-TOF MS analysis.
17. Remove the anionic exchange filtration plate from the vacuum manifold and add 100 µl of Wash Buffer 3 containing 100 mM sodium acetate (aq) and 0.1% (v/v) *n*-Octyl β-D-glucopyranoside (OGP) (pH 5) to each sample well.
18. Repeat steps 12–16 except using Wash Buffer 3 and that subsequent eluent should be labeled Fraction 3.
19. Remove the anionic exchange filtration plate from the vacuum manifold and add 100 µl of Wash Buffer 4 containing 100 mM sodium acetate (aq) and 0.1% (v/v) *n*-Octyl β-D-glucopyranoside (OGP) (pH 4) to each sample well.
20. Repeat steps 12–16 except using Wash Buffer 4 and that subsequent eluent should be labeled Fraction 4.
21. Remove the anionic exchange filtration plate from the vacuum manifold and add 100 µl of Wash Buffer 5 containing 50 mM sodium citrate (aq) and 0.1% (v/v) *n*-Octyl β-D-glucopyranoside (OGP) (pH 3) to each sample well.
22. Repeat steps 12–16 except using Wash Buffer 5 and that subsequent eluent should be labeled Fraction 5.
23. Remove the anionic exchange filtration plate from the vacuum manifold and add 100 µl of Wash Buffer 6 containing an aqueous solution of 33.3% (v/v) isopropanol, 16.7% (v/v) acetonitrile, and 0.1% (v/v) trifluoroacetic acid to each sample well.
24. Repeat steps 12–16 except using Wash Buffer 6 and that subsequent eluent should be labeled Organic Fraction.

25. This microscale anionic exchange protocol generates a total of six separate fractions (i.e. proteins eluted at pH 9, pH 7, pH 5, pH 4, pH 3, and the organic fraction) per original muscle lysate sample, each sample fraction being 200 µl in volume.

3.4. SELDI-TOF MS Analysis of Muscle Lysate Fractions

1. Twelve, eight spot CM-10 chips (a weak cationic exchange surface) were placed in a 96-well plate format bioprocessor unit (BioRad Laboratories).

2. Each well of the bioprocessor unit (the base of each well in the bioprocessor corresponds to a single spot on each chip) is washed with three changes of binding buffer consisting of 50 mM sodium acetate (aq) (pH 4) over a period of 10 min.

3. During each wash, the buffer is aspirated up and down in the well to ensure any air bubbles trapped in the wells of the bioprocessor are dislodged in order to allow adequate hydration and washing of the spot surface with binding buffer prior to sample application. This can also be achieved by centrifuging the bioprocesor unit in a table top centrifuge that allows centrifugation of 96-well plates.

4. Prior to sample application, individual muscle lysate fraction samples are diluted by a factor of 10 using binding buffer (i.e. 20 µl of sample is added to 180 µl of 50 mM sodium acetate (aq) – pH 4).

5. After removal of the binding buffer by aspiration from the bioprocessor wells, individual samples of diluted muscle lysate fractions are immediately aliquoted into individual wells of the bioprocessor unit.

6. Duplicate samples are aliquoted onto different chips to assess chip-to-chip variation in the protein profiles generated during analysis.

7. One spot per chip is reserved for a mixture of standard proteins consisting of a pooled serum sample to assess overall reproducibility within the analysis.

8. Samples are incubated in the bioprocessor unit at 4°C overnight with gently mixing to allow binding of proteins to the spot surface of the chip.

9. After protein binding overnight, each well of the bioprocessor unit is washed with three changes of fresh binding buffer over a period of 20 min, followed by one rapid wash with distilled water for a period of 10 s.

10. The chip arrays are then removed from bioprocessor unit and allowed to air dry.

11. Two microliters of energy-absorbing matrix (EAM), consisting of 12.5 mg/ml of sinapinic acid dissolved in solution of 0.5% TFA/50% acetonitrile (aq), are applied to the surface of each spot on each of the chip arrays.

12. The EAM is then allowed to air dry on the surface of each spot prior placing in the SELDI-TOF MS.

13. SELDI-TOF MS analysis is carried out to produce protein profiles contained in each of the muscle lysate fractions.

14. TOF MS conditions are set based upon each experiment with regard to laser intensity and analysis conditions. These conditions are arrived at by first determining the data acquisition conditions required to collect consistent protein profiles from the standard protein mixture samples located on each chip and then applying those data acquisition settings to all subsequent muscle lysate samples.

4. Data Analysis

The experimental approach detailed above was used to generate a series of protein profiles from control, atrophied, and "atrophy-rescued or DFS treated" rat soleus muscle samples. By comparing control soleus muscle protein profiles to those obtained from atrophied and "atrophy-rescued" muscle, we have detected several proteins which appear to be modulated during atrophy compared to that observed in control or "atrophy-rescued" muscle (Fig. 1). These data indicate that a subtractive protein-profiling approach using SELDI-TOF MS (combined with standard protein fractionation techniques) can be used to investigate the etiology of disuse atrophy in skeletal muscle. In addition, this approach can be used to identify proteins that are specifically involved in the control or modulation of the muscle atrophy response, rather than a systemic response to disuse, by comparing these protein changes across control, experimental and rescue conditions.

Fig. 1. Detection of protein changes associated with unloading-induced muscle atrophy using SELDI-TOF MS after fractionation of muscle lysates using microscale ion exchange chromatography. (a) Note the protein with a molecular mass of approximate 3,835 Da which is absent in control muscle, up-regulated in atrophied muscle and down-regulated in DFS-treated muscle (i.e. atrophy-rescued). These proteins were detected in the organic solvent fraction generated during anionic exchange separation of muscle lysates. (b) Note the protein with a molecular mass of approximate 4,120 Da which is absent in control muscle, up-regulated in atrophied muscle and down-regulated in DFS-treated muscle (i.e. atrophy-rescued). These proteins were detected in Fraction 5 (pH 3 buffer wash) generated during anionic exchange separation of muscle lysates. (c) Note the proteins with a molecular mass of approximate 43,800–44,400 Da which are present in both control and DFS-treated muscle, but down-regulated in atrophied muscle. These proteins were detected in Fraction 4 (pH 4 buffer wash) generated during anionic exchange separation of muscle lysates.

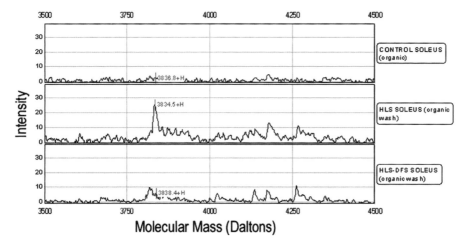

Panel A

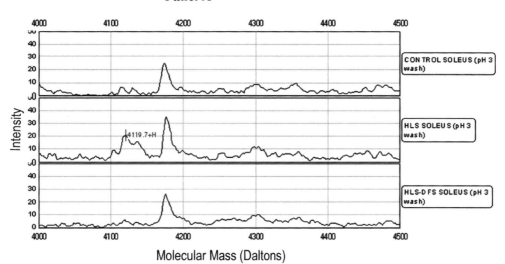

Panel B

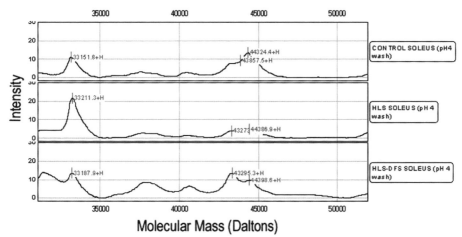

Panel C

5. Notes

As with all protein analysis techniques, care should be taken to minimize proteolytic activity at all stages of sample preparation. When collecting muscle for subsequent analysis, samples should be excised as soon as possible and immediately snap frozen in liquid nitrogen. Samples should not be subjected to repeated freeze-thaw cycles as this can lead to significant sample degradation. During muscle lysate preparation, all buffers should contain protein inhibitors and samples should be stored on ice whenever possible. Although the protocol detailed in this chapter uses manual techniques, the use of a robotic fluid handling system if available should be considered. This reduces "sample-to-sample" variations associated with the multiple reagent pipetting steps involved during microscale ion exchange chromatography and protein chip array preparation in the bioprocessor unit. To ensure accuracy and reproducibility when generating protein profiles utilizing the SELDI-TOF MS technology, TOF MS equipment should be regularly maintained and calibrated on a weekly basis. The inclusion of a consistent complex protein standard, such as a pooled serum sample, within and between experimental runs ensures that results obtained over time are comparable, and can be used to account for batch-to-batch variations between protein chip arrays of the same type.

Acknowledgments

The author would like to acknowledge Dr. Antonios Kyparos for providing the muscle samples used in this study. This study was facilitated by the existence of a Space Act Agreement between NASA-Johnson Space Center and the University of Houston which allowed access to the SELDI-TOF MS equipment located at NASA-JSC.

References

1. Clarke, M.S., *The effects of exercise on skeletal muscle in the aged.* J Musculoskelet Neuronal Interact, 2004. **4**(2): p. 175–8.
2. Giangregorio, L. and N. McCartney, *Bone loss and muscle atrophy in spinal cord injury: epidemiology, fracture prediction, and rehabilitation strategies.* J Spinal Cord Med, 2006. **29**(5): p. 489–500.
3. Ferrando, A.A., D. Paddon-Jones, and R.R. Wolfe, *Bed rest and myopathies.* Curr Opin Clin Nutr Metab Care, 2006. **9**(4): p. 410–5.
4. Argiles, J.M., S. Busquets, and F.J. Lopez-Soriano, *The pivotal role of cytokines in muscle wasting during cancer.* Int J Biochem Cell Biol, 2005. **37**(10): p. 2036–46.

5. Fitts, R.H., D.R. Riley, and J.J. Widrick, *Functional and structural adaptations of skeletal muscle to microgravity.* J Exp Biol, 2001. **204**(Pt 18): p. 3201–8.

6. Boonyarom, O. and K. Inui, *Atrophy and hypertrophy of skeletal muscles: structural and functional aspects.* Acta Physiol (Oxf), 2006. **188**(2): p. 77–89.

7. Baldwin, K.M., *Effects of altered loading states on muscle plasticity: what have we learned from rodents?* Med Sci Sports Exerc, 1996. **28**(10 Suppl): p. S101–6.

8. Adams, G.R., V.J. Caiozzo, and K.M. Baldwin, *Skeletal muscle unweighting: spaceflight and ground-based models.* J Appl Physiol, 2003. **95**(6): p. 2185–201.

9. Kyparos, A., et al., *Mechanical stimulation of the plantar foot surface attenuates soleus muscle atrophy induced by hindlimb unloading in rats.* J Appl Physiol, 2005. **99**(2): p. 739–46.

10. Jackman, R.W. and S.C. Kandarian, *The molecular basis of skeletal muscle atrophy.* Am J Physiol Cell Physiol, 2004. **287**(4): p. C834-43.

11. Musacchia, X.J., J.M. Steffen, and R.D. Fell, *Disuse atrophy of skeletal muscle: animal models.* Exerc Sport Sci Rev, 1988. **16**: p. 61–87.

12. Tesch, P.A., et al., *Effects of 17-day spaceflight on knee extensor muscle function and size.* Eur J Appl Physiol, 2005. **93**(4): p. 463–8.

13. Trappe, S., et al., *Human single muscle fibre function with 84 day bed-rest and resistance exercise.* J Physiol, 2004. **557**(Pt 2): p. 501–13.

14. Bamman, M.M., et al., *Impact of resistance exercise during bed rest on skeletal muscle sarcopenia and myosin isoform distribution.* J Appl Physiol, 1998. **84**(1): p. 157–63.

15. Bamman, M.M., et al., *Enhanced protein electrophoresis technique for separating human skeletal muscle myosin heavy chain isoforms.* Electrophoresis, 1999. **20**(3): p. 466–8.

16. Kandarian, S.C. and E.J. Stevenson, *Molecular events in skeletal muscle during disuse atrophy.* Exerc Sport Sci Rev, 2002. **30**(3): p. 111–6.

17. Zhang, P., X. Chen, and M. Fan, *Signaling mechanisms involved in disuse muscle atrophy.* Med Hypotheses, 2007. **69**(2): p. 310–21.

18. Haddad, F., K.M. Baldwin, and P.A. Tesch, *Pretranslational markers of contractile protein expression in human skeletal muscle: effect of limb unloading plus resistance exercise.* J Appl Physiol, 2005. **98**(1): p. 46–52.

19. Taylor, W.E., et al., *Alteration of gene expression profiles in skeletal muscle of rats exposed to microgravity during a spaceflight.* J Gravit Physiol, 2002. **9**(2): p. 61–70.

20. Lee, S.J. and A.C. McPherron, *Regulation of myostatin activity and muscle growth.* Proc Natl Acad Sci USA, 2001. **98**(16): p. 9306–11.

21. Wolfman, N.M., et al., *Activation of latent myostatin by the BMP-1/tolloid family of metalloproteinases.* Proc Natl Acad Sci USA, 2003. **100**(26): p. 15842–6.

22. Chen, Y.W., et al., *Transcriptional pathways associated with skeletal muscle disuse atrophy in humans.* Physiol Genomics, 2007. **31**(3): p. 510–20.

23. Wang, X., et al., *Mergla K+ channel induces skeletal muscle atrophy by activating the ubiquitin proteasome pathway.* Faseb J, 2006. **20**(9): p. 1531–3.

24. McKinnell, I.W. and M.A. Rudnicki, *Molecular mechanisms of muscle atrophy.* Cell, 2004. **119**(7): p. 907–10.

25. Raddatz, K., et al., *A proteome map of murine heart and skeletal muscle.* Proteomics, 2008. **8**(9): p. 1885–97.

26. Gelfi, C., et al., *The human muscle proteome in aging.* J Proteome Res, 2006. **5**(6): p. 1344–53.

27. Isfort, R.J., et al., *Proteomic analysis of the atrophying rat soleus muscle following denervation.* Electrophoresis, 2000. **21**(11): p. 2228–34.

28. Moriggi, M., et al., *A DIGE approach for the assessment of rat soleus muscle changes during unloading: effect of acetyl-L-carnitine supplementation.* Proteomics, 2008. **8**(17): p. 3588–604.

29. Clarke, M.S.F., S. Weinberger, and C.H. Clarke, *Detection of protein biomarkers during oxidative stress using Surface Enhanced Laser Desorption/Ionization-Time of Flight (SELDTOF) Mass Spectrometry.*, in *Oxidative Stress and Aging: Advances in Basic Science, Diagnostics and Intervention*, R.G.C.H. Rodriguez, Editor. 2002, World Scientific Publishing Co. p. 366–379.

30. Engwegen, J.Y., et al., *Clinical proteomics: searching for better tumour markers with SELDI-TOF mass spectrometry.* Trends Pharmacol Sci, 2006. **27**(5): p. 251–9.

31. Pisitkun, T., R. Johnstone, and M.A. Knepper, *Discovery of urinary biomarkers.* Mol Cell Proteomics, 2006. **5**(10): p. 1760–71.

32. Xiao, Z., et al., *Proteomic patterns: their potential for disease diagnosis.* Mol Cell Endocrinol, 2005. **230**(1–2): p. 95–106.

Chapter 11

Profiling Cervical Lavage Fluid by SELDI-TOF Mass Spectrometry

Adam Burgener

Abstract

The mucosal surface of the female genital tract is the first site of contact for many sexually transmitted infections and serves as the first layer of defense. This layer has components of both the innate and adaptive immune systems to protect against infections. For these reasons, this fluid is a major focus of study to understand the pathogenesis of different infectious diseases. Novel tools are available to allow for the analysis of the components of this mucosal layer, including the area of proteomics. The emergence of proteomics has allowed for the development of many types of platforms for protein profiling, including gel-based technologies (2-dimensional differential in-gel electrophoresis) and mass spectrometry-based techniques. SELDI-TOF, a mass spectrometry-based platform coupled to on-chip chromatographic separation, has been developed as a high-throughput technique to profile complex protein samples. This chapter will outline detailed methods to profile cervical mucosal samples by SELDI-TOF and can serve as a guideline for other types of mucosal samples.

Key words: SELDI-TOF, Cervical mucosa, Biomarkers, Female genital tract, Cervical lavage

1. Introduction

The mucosal layer of the lower female genital tract serves as the first layer of protection against infection from many infectious organisms. The fluid itself is secreted by epithelial cells in the vagina, and also other compartments that enter into the vagina, such as the endocervix and the oviduct. This fluid is host to many antimicrobial and antiviral components that are key components to innate defense (1–3). For these reasons, mucosal fluids are the focus of investigation by the infectious disease research community, including those in HIV/AIDS (4–7). The mucosa of the female genital tract is also studied for biomarkers of other conditions such as pre-term birth (8), intra-amniotic infection (9), and cancer (10). However, the complexity of mucosal fluids and its many components provides a challenge for researchers to study.

The field of proteomics has provided a plethora of techniques to examine biological fluids in a comprehensive manner. Many have been applied to examine cervical mucosal fluid, including traditional gel-based techniques, such as 2D-DIGE (two-dimensional differential in-gel electrophoresis), and more recently 2D-LC MS (two-dimensional liquid chromatography coupled to mass spectrometry) (5, 8, 9). However, some of the issues with these technologies is the low-throughput and the often necessity of pooling of individual samples prior to analysis.

SELDI-TOF technology offers an alternative approach to these methods. Its combination of high-throughput capacity and chromatographic separation makes it an ideal technology for profiling of cervical mucosa samples. The incorporation of chromatographic surfaces allows for the selection of different subsets of proteins, enhancing coverage, and simplifying downstream analysis. Indeed, studies using SELDI-TOF on cervical mucosa samples have yielded important discoveries, including those involved with resistance to HIV-1 infection (11), underscoring its applicability for protein profiling.

This chapter provides detailed methods and protocols for the analysis of cervical mucosal fluid by SELDI-TOF mass spectrometry. The focus is to provide suitable protocols for semi-quantitative comparisons between samples, but these methods can also be used as a guideline for basic profiling. Two different approaches for profiling cervical lavage are described, including profiling of neat lavage and the use of Proteominer™ beads as a pre-fractionation step prior to SELDI-TOF analysis to enhance overall protein coverage.

2. Materials

2.1. Materials for Profiling

1. Low retention Microcentrifuge tubes (1.5 ml) (see Note 1).
2. Low retention pipette tips (10 μl to 1.0 ml).
3. 1 M Sodium acetate, pH 4.0.
4. 0.1 M Sodium acetate buffer, pH 4.0.
5. 500 mM Tris–HCl, pH 8.0.
6. 50 mM Tris–HCl, pH 8.0.
7. 0.1 M sodium phosphate 0.5 M NaCl.
8. 100 mM Copper sulfate.
9. Micromix 5 or rotary shaker.
10. Microcentrifuge.

11. 250 µl pipette tips.
12. 20 µl pipette tips.
13. HPLC grade water.
14. Sinapinic acid (SPA).
15. Acetonitrile.
16. 1.0% Trifluoroacetic acid solution (1.0 ml TFA + 99 ml de-ionized H_2O).
17. CM10 ProteinChip arrays (BioRad).
18. IMAC30 ProteinChip arrays (BioRad).
19. Q10 ProteinChip arrays (BioRad).

2.2. Materials for Enrichment of Low Abundance Proteins Using ProteoMiner Beads

1. Proteominer™ Bead Columns (BioRad).
2. Elution reagent 1: 2.2 M thiourea, 7.7 M urea, 4.4% CHAPS.
3. Elution reagent 2: 9 M urea, 25 mM citric acid, 2% CHAPS, pH 3.5.
4. Elution reagent 3: 33.3% isopropyl alcohol, 16.7% acetonitrile, 0.5% TFA.
5. Wash buffer: 1× PBS (150 mM NaCL, 10 mM NaH_2PO_4).
6. Low retention pippette tips.
7. De-ionized water.

3. Methods

3.1. Sample Collection and Storage

The protocol to collect cervical lavage involves washing the endocervix with 2.0 ml of sterile 1× phosphate buffer saline (PBS) and collecting the lavage from the posterior fornix. Samples are then placed into a 15-ml conical tube, centrifuged to remove cellular debris (500 × g), and the supernatant stored in two 1.0 ml aliquots at −70°C until analysis.

3.2. Sample Preparation for Spotting onto ProteinChip Arrays

1. Samples should be thawed at 4°C (in the fridge, or on ice) just prior to analysis. Avoid extensive exposure of the sample to >4°C temperatures as sample degradation can occur.
2. Once thawed, samples should be re-centrifuged at 10,000 × g for 10 min at 4°C to remove any residual cell debris.
3. Remove supernatant and transfer into new 1 ml low retention microcentrifuge tube.
4. A total of 2.5 µg of protein is required per spot (100 µl of a 0.025 mg/ml sample is required) (see Note 2).

5. For samples with ≥0.25 mg/ml:
   (a) Add a total of 2.5 µg protein to fresh microcentrifuge tube.
   (b) Dilute to 100 µl with an appropriate binding buffer:
      - CM10 arrays : 0.1 M sodium acetate, pH 4.0.
      - Q10 arrays: 50 mM Tris–HCl, pH 8.0.
      - IMAC30 arrays: 0.1 M sodium phosphate 0.5 M NaCl.
   (c) For example, if sample protein concentration is 0.25 mg/ml, add 10 µl of sample + 90 µl of buffer.
6. For samples with <0.25 mg/ml:
   (a) To reduce the amount of dilution, use concentrated stock solutions to adjust the pH of the sample:
      - CM10 arrays: 1 M Sodium acetate, pH 4.0.
      - Q10 arrays: 500 mM Tris–HCl, pH 8.0.
   (b) Add 20 µl of 10× concentrated binding buffer to 180 µl of CVL sample.
   (c) Note that no adjustment is necessary for IMAC30 arrays since samples are collected in PBS.
   (d) Dilute to the sample to 0.025 mg/ml with appropriate binding buffer acetate buffer.
7. They should then be ready for spotting onto ProteinChip arrays.

3.3. Spotting Samples onto ProteinChip Arrays

1. Take out the appropriate buffer from the fridge allow it to warm to room temperature. (100 ml is enough to process 1 full bioprocessor of chips).
2. Place the ProteinChip arrays to be used into the Bioprocessor.
3. Pre-charge IMAC arrays with desired metal ion:
   (a) Pretreat the IMAC chips with 50 µl of 100 mM $CuSO_4$ to each well to load copper ions. Shake the bioprocessor for 5 min.
   (b) Remove the buffer and wash with 100 µl water per well. Remove the water and proceed to step 4.
4. Add 200 µl of appropriate binding buffer to each well. Binding buffer compositions include:
   (a) CM10 arrays: 0.1 M sodium acetate, pH 4.0.
   (b) Q10 arrays: 50 mM Tris–HCl, pH 8.0.
   (c) IMAC30 arrays: 0.1 M sodium phosphate 0.5 M NaCl.
5. Incubate for 5 min at room temperature with shaking (Micromix, setting 20/7 (see Note 3), or at 250 rpm with a rotary shaker).
6. Remove buffer from the wells.

7. Add 100 µl of the equilibrated sample (prepared in Subheading 3.2) to each well.
8. Incubate the bioprocessor with shaking for 30 min (Micromix, setting 20/7, or at 250 rpm with a rotary shaker).
9. Remove the samples from the wells and wash with 200 µl of binding buffer for 5 min, with shaking. Discard the wash.
10. Repeat wash two more times.
11. Remove the binding buffer and wash each well with 200 µl of HPLC grade H_2O. Shake for 5 min and repeat once more.
12. Remove the reservoir from the bioprocessor and allow the chips to dry at room temperature (typically 10–15 min).
13. During this time make the SPA solution. Add 0.5 ml of 100% acetonitrile and 0.5 ml of 1.0% TFA solution to a vial containing 10.0 mg of SPA. Vortex for 5 min, and centrifuge briefly prior to use.
14. Add 0.5 µl of SPA solution to each spot. Air dry for 5 min, and repeat with another 0.5 µl of SPA solution. Allow to air dry.
15. Analyze by SELDI-TOF.

Figure 1 shows a typical SELDI-TOF spectrum of a CVL sample profiled on a CM10 ProteinChip array. In this figure, both the low molecular weight (2–15 kDa) (top) and the high molecular weight range (15–100 kDa) (bottom) are shown, with the presence of many independent protein peaks. The most predominant peaks observed in CVL samples by SELDI-TOF are those in the low MW range, and an albumin peak at 66 kDa.

Figure 2 shows the same CVL sample profiled on CM10, Q10, and IMAC30 ProteinChip arrays. Comparable amounts of peaks are observed in each chip type and there is considerable overlap. Some of the peaks show up on all of the chromatographic surfaces, suggesting these proteins have cationic, anionic, and metal-affinity residues. A perfect example of this is the triplet peak starting at 3,440 Da.

It is common to observe significant variation in the protein concentration between individual samples and their corresponding mass spectra. This may be due to many factors. The first is the nature of the collection method itself as the washing of the ectocervix may not remove the same amount of mucosal fluid from each individual. Secondly, the female genital tract is continually shedding cells, which add cell debris into the mucosal milieu, increasing the protein concentration and adding variability in the spectra. To illustrate the type of variability one may encounter, Fig. 3 shows CVL samples from four different individuals, profiled on CM10 ProteinChip arrays (CM10) with the same amount of protein (2.5 µg) per spot. The type of variability between individuals is commonly observed and can be compensated for by screening a larger number of patient samples.

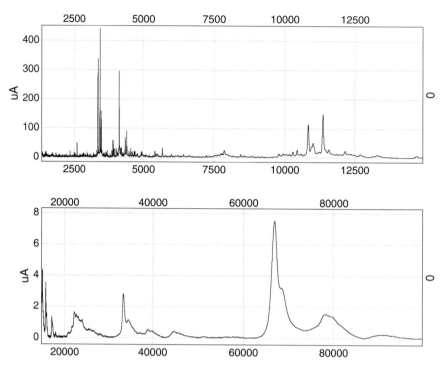

Fig. 1. *SELDI-TOF spectra of a representative CVL sample profiled on CM10 ProteinChip arrays.* 2.5 µg of a CVL standard was spotted onto a CM10 ProteinChip array.

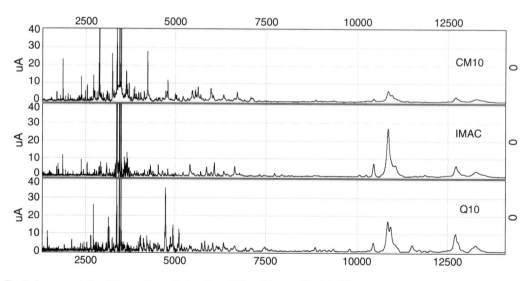

Fig. 2. *A representative CVL sample profiled on three different types of ProteinChip arrays.* 2.5 µg of a pooled CVL standard was spotted onto a CM10, IMAC30, and Q10 Proteinchip arrays.

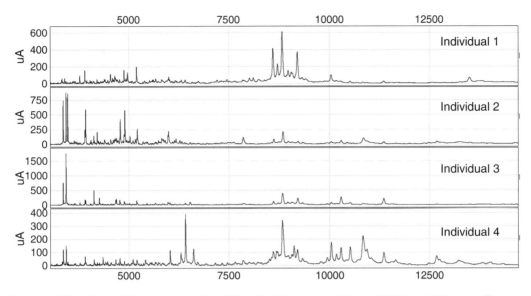

Fig. 3. *Variability of cervical lavage samples from different individuals as profiled on CM10 ProteinChip arrays.* The same amount of protein from each individual (2.5 μg) was spotted onto CM10 chips.

3.4. Proteominer Bead Fractionation Prior to SELDI Profiling of CVL Fluid

The study of proteins in complex biological samples continues to be a challenge for biomedical research. One major issue is the wide dynamic range of protein abundance that can span many orders of magnitude. Predominant protein species make the detection of low abundance proteins difficult as they overshadow their detection by mass spectrometry techniques. Cervical mucosal fluid is no different and is dominated by a few common protein species such as albumin and immunoglobulins (12).

As CVL samples typically contain a low amount of protein, it is desirable to utilize a method that could both concentrate the sample while removing high abundance proteins. A relatively new technology, termed Proteominer™ beads (PM-beads), has recently been introduced to the proteomics field as a promising fractionation method (13). Each bead is conjugated to a unique hexameric peptide sequence which creates a diverse library. This should represent enough sequences such that any protein in a complex proteome would have at least one ligand to bind. Low abundance proteins are concentrated on the beads while the levels of high abundance proteins are decreased as they saturate their corresponding ligands and are therefore washed through in the flow through. This resin is then washed and the captured proteins are eluted by chemical means. Proteominer beads have been utilized in other studies of complex biological proteomes, such as human urine (14) and human serum (15). In each case, the number of detectible protein species was increased significantly, by as much as 50–400%.

We have evaluated PM-bead technology to pre-fractionate cervical lavage prior to analysis with SELDI-TOF MS. We found that utilizing these beads increased both the number and intensity of detectible peaks in the sample by as much as 90% on CM10 Proteinchip arrays, while at the same time reducing high abundance peaks (such as albumin at 66 kDa and hemoglobin at 14 kDa). This may be due to the relative decrease in concentration of high abundance proteins, which are known to suppress signal from lower abundance proteins in the sample. This has been a consistent observation in other studies using Proteominer beads where reducing the dynamic range between protein species greatly enhanced peak intensity (13–15). We have evaluated two different elution strategies from PM-beads for CVL samples:

TUC Method

- Fraction 1 (2.2 M thiourea, 7.7 M urea, 4.4% CHAPS).
- Fraction 2 (9 M urea, 25 mM citric acid, 2% CHAPS, pH 3.5).
- Fraction 3 (33.3% isopropyl alcohol, 16.7% acetonitrile, 0.5% TFA).

U/CA Method

- Fraction 1: 9 M urea, 25 mM citric acid, 2% CHAPS pH 3.5.
- Fraction 2: 33.3% isopropyl alcohol, 16.7% acetonitrile, 0.5% TFA.

We found the TUC buffers, with the inclusion of Thiourea, significantly increased the elution efficiency from the beads over the U/CA buffers. Therefore for CVL profiling, we would recommend the TUC elution method for eluting samples from PM-beads.

1. Thaw samples on ice or at 4°C.
2. Equilibrate the column by adding 1 ml of Wash buffer to the beads and gently rotate for 10 min (rotate the column end to end).
3. Remove the caps, place column into a collection tube, and centrifuge at $1,000 \times g$ for 2 min. Discard the flow-through.
4. Adjust the protein concentration of sample to 0.25 mg/ml with PBS. Total volume should be at least 1.0 ml.
5. Load 1.0 ml of 0.25 mg/ml CVL fluid to the column.
6. Gently rotate the column for 2 h on rotary shaker at room temperature (rotate the column end to end).
7. Remove the caps and centrifuge at $1,000 \times g$ for 2 min. Keep the flow-through.

8. Replace bottom cap and add 1.0 ml of wash buffer. Rotate the column for 5 min.

9. Remove caps and place into collection tube and centrifuge at $1,000 \times g$ for 2 min. Keep flow-through on ice.

10. Repeat steps 7 and 8 two more times.

11. Replace bottom cap and add 80 µl of elution buffer. Add top cap and rotate column for 5 min.

12. Remove caps and place into collection tube. Centrifuge for 2 min at $1,000 \times g$.

13. Keep eluate on ice.

14. Repeat steps 10 and 11 and pool with the previous eluate.

15. Repeat steps 10–13 for Elution reagent 2 and Elution reagent 3, for a total of two washes each.

16. Profile eluates on ProteinChip arrays as described in Subheading 3.2, adding 90 µl of binding buffer and 10 µl of sample to each well.

4. Notes

1. Whenever possible utilize low-retention pipette tips and microcentrifuge tubes. These are typically siliconized and reduce the probability of protein loss due to protein-tube affinity. This is especially important when dealing with low-protein cervical mucosa samples.

2. Cervical mucosal samples vary greatly in their protein concentration. It has been our experience with the analysis of thousands of cervical lavage samples that the protein concentration can vary from as little as 0.00 mg/ml to >2.0 mg/ml. However, the average protein concentration is typically around 0.40 mg/ml. Samples with <0.05 mg/ml of protein will not be very useful for SELDI-TOF profiling and should not be used or should be concentrated prior to use.

3. The micromixer performance can vary with age of the instrument. A 20/7 setting is recommended for newer machines, but might have to be adjusted to allow for appropriate agitation. The buffer in the wells should be moving in a constant swirling motion during the shaking procedure.

References

1. Valore, E.V., et al., *Antimicrobial components of vaginal fluid*. Am J Obstet Gynecol, 2002. **187**(3): p. 561–8.
2. Venkataraman, N., et al., *Cationic polypeptides are required for anti-HIV-1 activity of human vaginal fluid*. J Immunol, 2005. **175**(11): p. 7560–7.
3. Hein, M., et al., *Antimicrobial factors in the cervical mucus plug*. Am J Obstet Gynecol, 2002. **187**(1): p. 137–44.
4. Horton, R.E., et al., *Cervical HIV-Specific IgA in a Population of Commercial Sex Workers Correlates with Repeated Exposure But Not Resistance to HIV*. AIDS Res Hum Retroviruses, 2008.
5. Burgener, A., et al., *Identification of differentially expressed proteins in the cervical mucosa of HIV-1-resistant sex workers*. J Proteome Res, 2008. **7**(10): p. 4446–54.
6. Iqbal, S.M., et al., *Elevated T cell counts and RANTES expression in the genital mucosa of HIV-1-resistant Kenyan commercial sex workers*. J Infect Dis, 2005. **192**(5): p. 728–38.
7. Kaul, R., et al., *HIV-1-specific mucosal IgA in a cohort of HIV-1-resistant Kenyan sex workers*. Aids, 1999. **13**(1): p. 23–9.
8. Pereira, L., et al., *Identification of novel protein biomarkers of preterm birth in human cervical-vaginal fluid*. J Proteome Res, 2007. **6**(4): p. 1269–76.
9. Gravett, M.G., et al., *Proteomic analysis of cervical-vaginal fluid: identification of novel biomarkers for detection of intra-amniotic infection*. J Proteome Res, 2007. **6**(1): p. 89–96.
10. Bae, S.M., et al., *Two-dimensional gel analysis of protein expression profile in squamous cervical cancer patients*. Gynecol Oncol, 2005. **99**(1): p. 26–35.
11. Iqbal, S.M., et al., *Elevated elafin/trappin-2 in the female genital tract is associated with protection against HIV acquisition*. Aids, 2009. **23**(13): p. 1669–77.
12. Dasari, S., et al., *Comprehensive proteomic analysis of human cervical-vaginal fluid*. J Proteome Res, 2007. **6**(4): p. 1258–68.
13. Thulasiraman, V., et al., *Reduction of the concentration difference of proteins in biological liquids using a library of combinatorial ligands*. Electrophoresis, 2005. **26**(18): p. 3561–71.
14. Castagna, A., et al., *Exploring the hidden human urinary proteome via ligand library beads*. J Proteome Res, 2005. **4**(6): p. 1917–30.
15. Sennels, L., et al., *Proteomic Analysis of Human Blood Serum Using Peptide Library Beads*. J Proteome Res, 2007.

Chapter 12

Isolation and Proteomic Analysis of Platelets by SELDI-TOF MS

Sean R. Downing and Giannoula L. Klement

Abstract

Many growth factors, leukotrines, and biological ligands are not circulating free in plasma or serum, except in the case of late or disseminated disease. During early tumor growth and angiogenesis, platelets actively and selectively sequester regulators of angiogenesis and, as such, the platelet protein content can be used as a marker of early tumor growth or angiogenesis. With the recent increase in the clinical use of biologic modifiers in cancer and chronic disease therapy, the search for markers of early disease, therapeutic response, and/or recurrence has suggested that analysis of platelet proteins may be more relevant and accurate. We provide a guideline for the proteomic analysis of platelet proteome, placing specific emphasis on angiogenesis regulators, even though other platelet proteins may serve as markers of disease in the future. The analysis of serum/plasma has been fraught with difficulties because of the extraordinarily large number of proteins and because some of the proteins are contained in extraordinarily large amounts, masking the less abundant proteins. Thus, platelets may provide a much more biologically relevant analyte for biomarker discovery.

Key words: Platelet, Platelet proteomics, Angiogenesis, Markers of tumor growth, Vascular endothelial growth factor (VEGF), Basic fibroblast growth factor (bFGF), Platelet-derived growth factor (PDGF)

Abbreviations

Coag	Coagulation, referring to a test used to measure coagulation time of whole blood
EGF	Epidermal growth factor
Fbgn	Fibrinogen
bFGF	Basic fibroblast growth factor
PGE	Prostaglandin
PRP	Platelet rich plasma
PDGF	Platelet derived growth factor
PPP	Platelet poor plasma
SELDI-TOF MS	Surface enhanced laser desorption/ionization – time-of-flight mass spectrometry
SPA	Sinapinic acid
TFA	Trifluoroacetic acid
VEGF	Vascular endothelial growth factor

1. Introduction

The introduction of biologic modifiers into cancer therapy, treatment of chronic inflammatory disorders, and acute care has led to an intense search for circulating biomarkers of disease progression and therapeutic response. Considerable effort is currently focused on methods for early cancer detection, including those involving: detection of specific proteins or proteomic profiles in the serum (1–4), DNA in stool samples (5–8), and gene expression profiles in lesional biopsies (9–12). The realization that angiogenesis is a critical part of tumor progression of solid (13) and liquid (14, 15) tumors, led over the past few decades to numerous attempts to correlate plasma and serum levels of angiogenesis-regulatory proteins with disease progression (16, 17). While helpful in the identification of patients with disseminated disease, the reliability of serum and plasma levels of vascular endothelial growth factor (VEGF), basic fibroblast growth factor (bFGF), epidermal growth factor (EGF), platelet-derived growth factor (PDGF), or other angiogenesis related proteins in early-stage tumors remains uncertain (18, 19). The absence of biological ligands from plasma/serum is most likely due to the short half life of these proteins and the evolutionary disadvantage of retaining them free in circulation as many of them are toxic in high concentrations. High circulating levels of proteins that stimulate angiogenesis, such as VEGF or bFGF, have been shown to result in severe paraneoplastic syndrome (20–22). This syndrome, characterized by tachycardia, hypertension, nervousness, cachexia, and early death, has been well documented in disseminated malignancies, early occult neoplasms, as well as in severe inflammatory conditions and burns.

Early cancer invasion and injury are dependent on the spatial and temporal regulation of a local release of the relevant growth factors, cytokines, and other angiogenesis regulators. Similarly, platelets are crucial for the homing of endothelial cell progenitors (23) and inflammatory cells, thus regulating tissue re-generation and maintenance. The evolutionary necessity to localize inflammatory effectors and angiogenesis regulators also explains why we have not been able to detect elevated plasma/serum levels of these proteins during early disease initiation; and why we could not use the circulating levels of angiogenesis regulators as markers of early disease, disease recurrence, or progression. It has always been appreciated that the local concentration of angiogenesis regulators significantly exceeds that of plasma or serum (24) (Fig. 1), but the vehicle for the delivery of angiogenesis regulators to sites of activated endothelium within an early tumor, atherosclerotic lesion, wound, or a plague of endometriosis has only been described recently (25, 26).

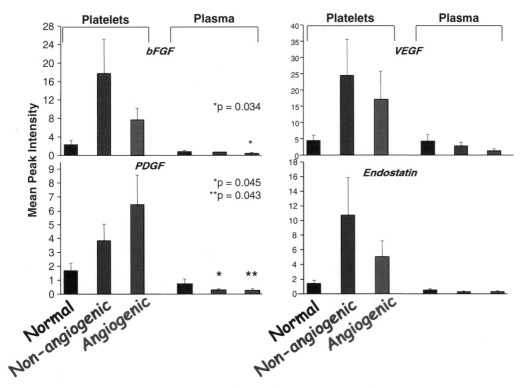

Fig. 1. Graphical representation of protein levels of angiogenesis regulators in plasma and platelets. We present four angiogenic regulators and their respective pattern of expression. The levels of pro-angiogenic growth factors (VEGF, bFGF, and PDGF) tend to be higher in the platelets, but not plasma, of tumor-bearing mice as compared to tumor-free sham operated mice. Although at 30 days nonangiogenic tumors are ~100 times smaller than their angiogenic counterparts, the platelet proteome in mice bearing nonangiogenic tumors shows detectable differences in the levels of angiogenesis regulators. In mice with nonangiogenic dormant tumors, the elevation of positive angiogenesis regulators tends to be counterbalanced by an elevation of the endogenous inhibitor endostatin. In contrast, in mice bearing the angiogenic clone, (red) platelets show a decrease in negative angiogenesis regulators such as endostatin. This suggests that, in the "angiogenic" tumors, the overall balance of angiogenesis regulators may be tipped toward a more pro-angiogenic phenotype in a manner reflected on the platelet proteome. Of note is the observation that all of the proteins are found at much higher concentrations in the platelets when compared to the plasma. The experiment was repeated on two separate occasions with five mice per each experiment, and the graph represents means ± standard error of means. This research was originally published in *Blood*. Klement GL, Yip TT, Cassiola F, Kikuchi L, et al. Platelets actively sequester angiogenesis regulators. *Blood*. 2009; 113:2835–2842. © American Society of Hematology.

We recently reported that platelets contain numerous proteins that regulate angiogenesis (27–29). Specifically, we had shown that the accumulation of angiogenesis regulators in platelets of animals bearing malignant tumors significantly exceeds their concentration in plasma or serum, as well as, their levels in platelets from non-tumor-bearing animals. This process is selective, as platelets do not take up a proportional amount of other plasma proteins (e.g. albumin), even though these may be present at higher concentrations (27, 30). We also found that VEGF-enriched Matrigel pellets implanted subcutaneously into mice, or the minute quantities of

VEGF secreted by microscopic subcutaneous tumors (0.5–1 mm^3), resulted in an elevation of VEGF levels in platelets, without any changes in plasma levels of the protein (27). The profile of other angiogenesis regulatory proteins (e.g. PDGF, bFGF) sequestered by platelets also reflected the presence of tumors in vivo before they were macroscopically evident. The ability of platelets to selectively take up angiogenesis regulators in cancer-bearing hosts may have implications for the diagnosis and management of many angiogenesis-related diseases, and provide a guide for anti-angiogenic therapies.

With the appreciation that platelets sequester, transport, and deliver angiogenesis regulators, the methodology for platelet isolation, purification, and analysis has gained in importance. In fact, with the increasing use of anti-angiogenic agents, growth factors, and other biologics, it is likely that the detection and quantification of relevant platelet proteins will continue to gain in importance. In addition, searches for protein biomarkers within the serum or plasma are hindered by the sheer number of proteins found within the blood. These proteins exist over an extremely large dynamic range spanning over 10 orders of magnitude (31), making the search for proteins in the plasma very difficult. A particular drawback has been the presence in serum/plasma of large amounts of albumin and immunoglobulins. These proteins, while present in platelets, are not actively sequestered by them and the levels of actively sequestered proteins are far easier to detect in platelets.

SELDI-TOF MS offers many advantages in the analysis of platelets over other methods of biomarker discovery because it allows you to quickly analyze sample sizes that are statistically relevant. Platelets are rich in proteins that regulate angiogenesis. Many of these proteins, if not all, are cationic with a pI > 8. The standard anion exchange fractionation procedure for SELDI-TOF MS separates these angiogenesis-regulating proteins from the bulk of proteins contained within platelets, especially albumin. The ability to analyze the "platelet angiogenesis proteome" partially purified from the remainder of the platelet proteome may provide distinct advantages in the search for novel biomarkers of disease and/or therapeutic response. While SELDI-TOF MS does not give you protein identification up front, purification and identification on the back-end of sample analysis is routine for many research laboratories.

2. Materials

2.1. Human Platelets

Human blood should only be drawn by a phlebotomist or trained medical professional. If the draw occurs within a phlebotomy clinic, or other medical professional office, no additional materials should be required for platelet acquisition other than routine laboratory supplies. However, if one is asked to provide the appropriate blood collection tubes, citrated tubes routinely used for coagulation (coag) studies may be purchased (see Note 1).

1. BD Vacutainer® (buffered sodium citrate: (9NC) blood collection tubes) (BD Biosciences, San Jose, CA).

2.2. Mouse Platelets

Outside of routine animal care facilities and standard surgical procedures and equipment, the following materials are required for mouse blood withdrawal:

1. Sterile sodium citrate buffer (105 mM, pH 5), stored at room temperature.
2. Needles (23–21 gauge) (BD Biosciences).
3. Syringes (1 cc) (BD Biosciences).

2.3. Fractionation

Fractionation of samples by anionic exchange utilizing a step-wise pH gradient reduces the overall complexity of a protein mixture thus making it easier to observe potential changes in the concentration of less abundant proteins. Samples need not necessarily be fractionated and can be run neat; however, this may reduce the likelihood of finding meaningful differences.

1. 96-well plates (v-bottom, low protein binding preferred).
2. Aluminum plate sealers (preferable to plastic because aluminum withstands storage at –80°C better and because the aluminum plate sealers can be punctured by a pipette tip whereas the plastic cannot).

The following materials need to be obtained or prepared if a ProteinChip Serum Fraction Kit (Bio-Rad Laboratories, Inc., Hercules CA) will not be used for fractionation.

1. Anion exchange Q ceramic HyperD F sorbent, 1.
2. 96-well filtration plate.
3. 9 M urea, 2% CHAPS, 50 mM Tris–HCl, pH 9 (U9 buffer).
4. 50 mM Tris–HCl, 0.1% OGP, pH 9 (wash buffer 1).
5. 50 mM HEPES, 0.1% OGP, pH 7 (wash buffer 2).
6. 100 mM sodium acetate, 0.1% OGP, pH 5 (wash buffer 3).
7. 100 mM sodium acetate, 0.1% OGP, pH 4 (wash buffer 4).
8. 100 mM sodium acetate, 0.1% OGP, pH 3 (wash buffer 5).
9. 33.3% isopropanol, 16.7% acetonitrile, 0.1% trifluoroacetic acid (TFA) (wash buffer 6).

2.4. Profiling

Currently, ProteinChip Arrays can only be purchased from Bio-Rad as it is proprietary technology. We highly recommend that all binding buffers also be purchased from Bio-Rad to avoid any quality control issues that may arise from researchers preparing their own buffers. While many chip chemistries are available, with the associated binding and wash buffers, we will concentrate on CM10 ProteinChip Arrays.

1. CM10 ProteinChip Arrays (Bio-Rad).
2. CM10 Low-Stringency Binding Buffer (Bio-Rad).
3. Acetonitrile.
4. 0.1% TFA.
5. Sinapinic acid (SPA).

3. Methods

The focus of this section is the collection and isolation of platelets from mice and humans as these two species offer examples of platelet collection from small and large organisms. The methods presented can be extrapolated to other experimental animal models. Platelets are delicate and, as such, a caveat for working with platelets is their sensitivity to external conditions and surfaces. Precautions need to be taken during collection and isolation to avoid platelet activation. Platelet activation causes changes in platelet morphology that ultimately lead to alterations in protein levels due to release of granular content (Fig. 2) (32). Due to the release of proteins during platelet contraction, proteomic analysis of platelets may not reflect the biological phenomenon being studied if care is not taken to prevent activation (see Note 2).

Platelet activation can be prevented by two means: (1) pharmacological pretreatment of the platelet preparations with calcium antagonist (verapamil), cyclooxygenase inhibitors (prostaglandin-2 (PGE_2)), or inhibiting fibrinogen (Fbgn) receptor activation using Fbgn binding peptide Gly-Pro-Arg-Pro (GPRP) (33–35) and (2) meticulous avoidance of activating surfaces and conditions. In deciding which pharmacological agent to use for the prevention of platelet activation, it is imperative that one considers their mechanism of action so as to not produce interference with the pathways of interest (see Note 3). An additional caveat to pharmacological pretreatment of platelets is the notion that the protein profile may change upon treatment. We have found that pretreatment with prostaglandin E significantly alters the level of a protein with an approximate mass of 14,620 Da (Fig. 3). To avoid activating surfaces and conditions, it is important to avoid: glass and polystyrene plastic surfaces, changes in temperature, dust, alterations in pH, shearing forces, and prolonged incubations without gentle agitation. In particular, putting whole blood or platelet preps on ice will

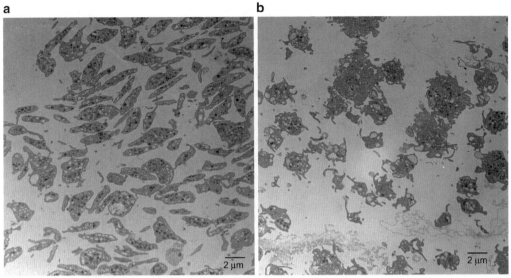

Fig. 2. Morphological changes in platelets due to activation. In a resting state, platelets are discoid in shape with two types of granules, alpha and dense granules (**a**). Platelets also contain lysosomes and mitochondria. The membrane of the platelet is interspersed with open canallicular system (OCS), which represents direct communications of the cytoplasm with the exterior of the platelet. This tenuous balance of the platelet membrane is crucial in the activation of platelets, which results in retraction of the granular content to the center of the platelet, widening of the OCS, and extension of filopodia, lamellipodia, and pseudopodia (**b**). Image is courtesy of Dr. Flavia Cassiola (27).

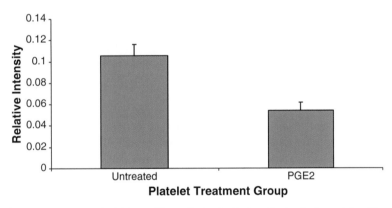

Fig. 3. Pretreatment of platelets with prostaglandin E alters the level of a protein with a mass of 14,620 Da. Platelets in PRP were incubated with PGE_2 for 1 h at 37°C or untreated, but given vehicle control. Following incubation, platelets were isolated, lysed, and analyzed by SELDI-TOF MS. Only the peak with a mass of 14,620 was found to be significantly altered by treatment with PGE_2 (p-value < 0.05). This is an important observation as pharmacological pretreatment of platelets to prevent activation may alter the protein content of the platelets.

cause irreversible damage to platelets and thus the proteomic profile. Pharmacological and methodological prevention of platelet activation should produce samples conducive to valid analysis of the protein content of platelets. We recommend that careful methodological protocols be employed and that pharmacological methods be used sparingly as the effects of their use on platelet protein content is not fully understood.

3.1. Platelet Collection

3.1.1. Collection of Human Platelets

The collection of human platelets for platelet aggregation and coagulation studies has been well established in the majority of clinical laboratories around the world, and as such does not require a detailed description. Briefly, platelets are a component of whole blood and can be readily obtained from a routine blood draw. However, unlike other blood components, the collection of platelets requires careful management of calcium chelation by ensuring that the blood draw maintains a volume per volume solution of 10% sodium citrate. Collection of human blood should be performed by a trained phlebotomist or appropriate medical professional, into standard blue top coagulation Vacutainers® containing citrate and filled to the top in order to preserve the appropriate citrate dilution. Use of a syringe or long tubing should be avoided in order to avoid platelet activation. A large boar needle is preferred to avoid sheering resulting in hemolysis and platelet activation (see Note 4). It is desirable to process whole blood samples immediately, or within 3 h of blood draw as long as the collection tubes are placed on a gentle rocker for that time period. Once platelet-rich plasma is isolated, it can be preserved on a gentle rocker for up to 3 days. Platelets react very quickly to the changes in sample pH and as such the quality of the sample decreases with prolonged metabolic activity that accompanies delays in processing. Unlike situations where cooling of the sample may decrease the metabolic activity, platelets should never be placed on ice, or in a cold environment, as this results in platelet lysis. Room temperature will suffice for most platelet applications although 37°C would be optimal.

3.1.2. Collection of Mouse Platelets

The collection of mouse platelets varies significantly from laboratory to laboratory and this variation may be the source of the variability of results seen in the literature. While mouse platelets are thought to be less sensitive to activation and can be *spin washed* in contrast to human platelets, all of the aforementioned rules for the collection of human platelets should be observed (see Note 5). To summarize, the steps involved in collection of platelets from animal experimental subjects are as follows:

1. Prepare a 105 mM solution of sterile sodium citrate buffer, pH 5.
2. Prepare a 1 cc syringe per mouse containing 100 µL of the sodium citrate buffer. This amount of buffer will allow for a final concentration of 10% volume per volume dilution of the citrate as long as the syringe is filled with blood to a volume of 1 mL.

3. Of the three methods to obtain blood from mice (orbital bleed, tail vein bleed, and direct cardiac puncture bleed), only the direct cardiac puncture provides sufficient flow and minimization of the shearing forces that may disrupt or activate platelets (see Note 6). Therefore, blood should be collected via a terminal cardiac bleed, under anesthesia, using a 23–21 gauge needle directly attached to the 1 cc, pre-citrated syringe.

4. Remove the needle from the syringe and place the whole blood specimen into a 1.5 mL polypropylene tube.

5. Place tube on a gentle rocker at room temperature or 37°C to avoid platelet activation.

3.2. Platelet Isolation

Human blood contains approximately 100,000–400,000 platelets per microliter. As long as the appropriate precautions are taken, and those platelet counts are preserved, a platelet pellet obtained from 1 mL of platelet-rich plasma (PRP) will contain ~1–1.5 mg of total protein (data not shown).

3.2.1. Isolation of Human Platelets

1. Centrifuge the whole blood collected in Subheading 3.1.1 for 20 min at $150 \times g$ using a swinging-bucket rotor at room temperature or 37°C. (Do not use a refrigerated centrifuge and do not place samples on ice).

2. Immediately following centrifugation, transfer the top phase, this is the platelet-rich plasma (PRP) (Fig. 4), to a fresh polypropylene tube. Care must be taken to avoid the buffy coat as this layer contains white blood cells that are nucleated and have a much greater protein content that will skew the proteomic analysis of platelets. For a standard 5 mL Vacutainer®, 5 mL of whole blood should yield ~2.5 mL of PRP (see Notes 7 and 8).

3. PRP should be divided into 1 mL aliquots, in order to facilitate standardization of the protein levels to the original platelet count. The goal is to have quantities of protein ultimately expressed as microgram of identified, differentially expressed protein per 100,000 platelets (see Note 9). A practical suggestion is to have this 1 mL aliquot in 1.5 mL polypropylene tubes labeled as "platelets" as this tube will end up containing the platelet pellet.

4. The 1.5 mL tubes containing the PRP are then centrifuged for 10 min at $900 \times g$ at room temperature or 37°C to pellet the platelets. Please note, centrifuging the platelets at higher g-forces will cause activation due to the collision of platelets with the bottom of the tube.

5. Following centrifugation of the platelets, the plasma (aqueous phase) is transferred to a fresh 1.5 mL polypropylene tube labeled "plasma." These tubes now contain platelet poor plasma (PPP).

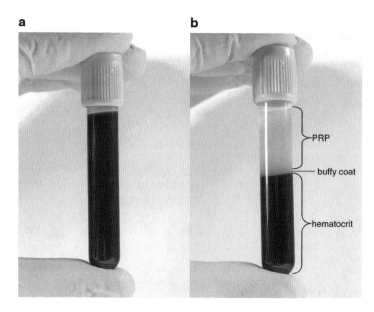

Fig. 4. Diagram of whole blood fractionation following centrifugation. Immediately following blood draw, the entire solution within the tube is homogenous and uniform in color (**a**). Following the first centrifugation at 150 × g, the whole blood separated into three distinct layers: PRP (*top layer*), buffy coat (*middle layer*), and the hematocrit (*bottom layer*). The hematocrit contains red blood cells, the buffy coat contains white blood cells, while the platelets are contained within the PRP. When removing the PRP, care should be taken to ensure that none of the buffy coat is taken up.

6. After removal of the PPP, residual plasma may still be present on the platelet pellet and tube wall. An additional step should be employed to remove this proteinacious material using non-fibrous filter paper to blot up the residuum. Note that plasma protein concentrations (in the 50 mg/mL range) will mask many of the less-abundant proteins in platelets unless this step is taken (see Note 9). Care should be taken to not disturb the platelet pellet.

7. Store plasma and platelet pellets at −80°C until further protein analysis.

3.2.2. Isolation of Mouse Platelets

Mouse blood contains approximately 500,000–1,000,000 platelets per microliter (36). Given that a smaller volume of whole blood is obtained, platelet pellets from mice are often difficult to observe upon centrifugation. Care should be taken to standardize the position of platelet pellets within microfuge tubes by always centrifuging tubes with the tabs of all tubes facing the same direction. Mouse platelets can be processed in a similar manner as human platelets; however, there are some differences in the isolation method that are worth noting.

1. Centrifuge whole blood in 1.5 mL tubes at $150 \times g$ for 30 min at room temperature or 37°C.

2. Immediately following centrifugation, remove the top phase and transfer to a fresh 1.5 mL tube labeled "platelets." The fresh tube now contains PRP. Care should be taken to avoid the buffy coat and the hematocrit layers as they contain white and red blood cells, respectively, and will contaminate the proteomic analysis. The original ~1 mL of whole blood obtained from a mouse should yield ~400–500 μL of PRP.

3. Centrifuge the tubes containing the PRP at $900 \times g$ for 30 min at room temperature or 37°C to pellet the platelets.

4. Remove the plasma and transfer to a new 1.5 mL tube. Label this tube "plasma" as it now contains the PPP.

5. Store plasma and platelet pellets at −80°C until further analysis.

3.3. Fractionation of Platelet Samples for SELDI-TOF MS

Standard SELDI-TOF procedures are utilized during platelet analysis with minor deviations from the manufacturer's protocols. Please reference all protocols from Bio-Rad Laboratories, Inc. for starting conditions and materials.

1. For analysis, lyse a pellet of human platelets obtained from 1 mL of PRP in 100 μL of U9 buffer (9 M urea, 2% CHAPS, 50 mM Tris–HCl, pH 9) and incubate at room temperature for at least 30 min. Mouse platelets obtained from ~0.5 mL of PRP should be lysed in 50 μL of U9 and incubated for at least 30 min. Platelet-poor plasma should be prepared according to standard SELDI-TOF methods for the analysis of serum. Optimally, 50 μL of plasma is incubated with 50 μL U9 and incubated for 30 min.

2. Following incubation in U9, add 100 μL U1 buffer (1 M urea, U9 buffer diluted 1:9) to the platelet lysate obtained from human platelets. For mouse platelets, 50 μL of U1 should be added. Add 100 μL U1 to denatured plasma samples.

3. Centrifuge samples at $13,000 \times g$ for 1 min to pellet any solid material such as cell membranes that have not dissolved.

4. One hundred microliters of the diluted human platelet lysate should be loaded onto a standard SELDI-TOF anionic exchange column for fractionation (rehydrated anion exchange Q ceramic HyperD F sorbent, 1). The remaining 100 μL can be used later for protein identification or immunoassays. For mouse platelets, the entire 100 μL of dilute platelet lysate should be used for fractionation (see Note 10). Diluted plasma from both species should be added at 100 μL.

5. Place the fractionation plate onto a shaker for 60 min at room temperature.

6. Place the fractionation plate onto a collection plate and use a vacuum manifold to draw the solution through the filer on the bottom of the fractionation plate into the collection plate.

Alternatively, the fractionation and collection plates can be placed into a swing-bucket centrifuge at 200–400 × g for 1 min (speed will vary depending on how viscous the sample is and how well the resin has been mixed).

7. Add 100 µL of wash buffer 1 (50 mM Tris–HCL, 0.1% OGP, pH 9) to the column and incubate with shaking for 10 min at room temperature.

8. Repeat the filtration step by either vacuum manifold or centrifugation collecting the flow through into the previous collection plate. The collection plate now contains Fraction 1 and can be stored at −80°C until analyzed by SELDI-TOF.

9. Add 100 µL wash buffer 2 (50 mM HEPES, 0.1% OGP, pH 7) to the column and incubate with shaking for 10 min at room temperature.

10. Place fractionation plate onto a new collection plate and repeat filtration step by either vacuum manifold of centrifugation.

11. Add another 100 µL of wash buffer 2 to the column, incubate with shaking for 10 min at room temperature, and collect into the collection plate from step 10. This is now Fraction 2 that can be stored at −80°C until analyzed.

12. Repeat steps 9–11 with wash buffers 3–6 to generate Fractions 3–6. Wash buffer 3 contains 100 mM sodium acetate and 0.1% OGP at pH 5 (wash buffers 4 and 5 are similar with pH 4 and 3, respectively). Wash buffer 6 contains 33.3% isopropanol, 16.7% acetonitrile, and 0.1% TFA.

13. For SELDI-TOF analysis, 50 µL of a fraction is added to 100 µL of the appropriate binding buffer for each chip chemistry to be utilized. Laser intensity needs to be determined empirically for each SELDI-TOF machine as this will vary depending on laser strength and age of the laser and detector. A typical spectrum from Fraction 1 on CM10 is shown in Fig. 5.

3.4. Profiling

This section concentrates on the use of CM10 ProteinChip Arrays. There are other chemistries available such as Immobilized Metal Affinity Capture (IMAC30) and Hydrophobic (H50) and the corresponding buffers, available from Bio-Rad, should be used in conjunction with the other chemistries.

1. Equilibrate CM10 ProteinChip Arrays with 150 µL CM10 Low Stringency Buffer for 5 min with vigorous shaking at room temperature.

2. Remove buffer and add 100 µL CM10 Low Stringency Buffer to the arrays.

3. Add 50 µL of fractionated sample to the arrays and incubate for 30 min with vigorous shaking at room temperature.

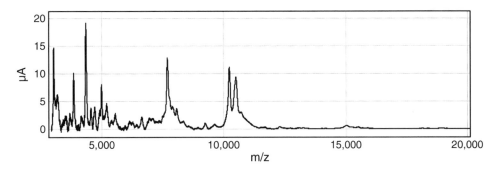

Fig. 5. Typical spectrum for Fraction 1 of a platelet lysate on CM10 surface arrays. A low laser energy was used to generate the spectrum resulting in mostly low molecular weight proteins being detected. One of the most abundant proteins in platelets is platelet factor 4 (PF-4) and this polypeptide is observed as the peak at approximately 7.8 kDa.

4. Remove samples from arrays.

5. Wash arrays three times with 150 µL CM10 Low Stringency Buffer for 5 min each wash with vigorous shaking at room temperature.

6. Quickly rinse arrays twice with 150 µL deionized water and allow to air dry.

7. Apply 1 µL of SPA to the arrays and allow to air dry.

8. Repeat step 7.

9. Read arrays on a SELDI-TOF MS at settings that are optimal for your samples and machine.

4. Notes

1. Collection for a coag study is a routine procedure performed in phlebotomy clinics. The procedure involves veni-puncture using a large bore needle and collection of blood directly into the citrated tube without the use of a syringe or excess tubing. For best results, a pre-clear of 1–2 mL of blood prior to sample withdrawal should be performed.

2. During activation, the platelet membrane retracts while extending pseudopodia and filopodia, the granular content is centered, and the open canalicular system enlarges (Fig. 2). These morphological and physiological alterations are associated with degranulation of the dense granules, calcium flux, and ATP release. Contrary to common belief, the release of proteins from alpha granules may be under different, as of yet undescribed, controls (27).

3. When choosing to pharmacologically inhibit platelet activation, care should be taken to ensure that the compound does not

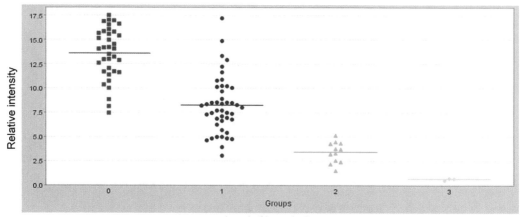

Fig. 6. Hemolysis alters the protein profile on CM10 arrays. Plasma samples obtained from 96 prostate cancer patients were segregated into four groups based on their level of hemolysis. Hemolysis was detected by eye and determined by the level of "redness." Group 0 indicates samples in which no detectable red color was observable and hence contains no hemolysis (36 samples). Group 1 contains samples that were slightly "pink" (45 samples), group 2 contains samples that were more orange in color (12 samples), and group 3 contains samples that were "blood red" (3 samples). A protein with an approximate mass of 66 kDa (assumed to be serum albumin) was significantly decreased with increasing hemolysis levels. This most likely was the result of increased hemoglobin levels out-competing albumin for binding sites on the CM10-binding surface.

interfere with the process under investigation. For example, the use of prostaglandin to inhibit platelet activation may not be conducive to studying the affects of aspirin on the platelet proteome.

4. Conditions that produce hemolysis in the sample should be avoided as anything that disrupts the plasma membrane of an erythrocyte will most likely affect the membrane of a platelet in a similar fashion. Furthermore, we have found that hemolysis has a profound impact on the proteins detected by SELDI-TOF MS. In our studies, we categorized plasma samples, by eye, from prostate cancer patients into four levels of hemolysis ranging from no hemolysis to extreme hemolysis (samples were blood red), with two intermediate levels of hemolysis being a slight pink color to a darker, slightly red color. In addition to the hemoglobin peaks found at ~15.2 and 15.9 kDa, several other peaks were able to distinguish between samples based on hemolysis levels (Fig. 6). Because hemolysis levels alter the protein content of the plasma, and potentially platelets, and this alteration of protein content leads to a differential protein detection by SELDI-TOF MS, we recommend that samples with hemolysis be discarded and not used for biomarker discovery or validation.

5. We do not recommend washing platelets. In our studies, we have found that washing of platelets is an uncontrollable variable

that causes partial activation of platelets, leading to an alteration of protein content, and results in a loss of approximately half of the platelets within the sample; making it difficult to normalize to platelet counts obtained from the original sample. However, it is important to remove as much as of the plasma as possible as described in Subheading 3.2 in order to avoid masking less abundant proteins.

6. We recommend terminal cardiac puncture bleeds because it provides sufficient flow and minimizes traumatic collection and hemolysis. Furthermore, terminal cardiac bleed allows for the collection of the greatest amount of blood when compared to orbital and tail vein blood draws. A typical blood volume for a mouse is 2 mL; terminal cardiac puncture generally provides 1 mL of blood while orbital and tail vein bleeds yield approximately 200 µL. In addition, orbital and tail vein bleeds produce tissue factor contact from the wound with the collected specimen. The production of a thrombus will alter the platelet number within the plasma and cause activation of platelets resulting in altered granular content of the platelets.

7. Following centrifugation, whole blood will separate into three phases: plasma (top phase), buffy coat (middle phase), and the hematocrit (bottom phase) (Fig. 4). In many cases, it is difficult to determine where the boundary between the plasma and buffy coat occurs. When removing the platelet-rich plasma from the separated whole blood, it is preferable to leave some of the platelet-rich plasma behind to ensure that none of the buffy coat is intermingled with the platelet-rich plasma as the buffy coat contains white blood cells.

8. The amount of plasma obtained from whole blood should be approximately half as long as the hematocrit is normal, i.e., 5 mL of whole blood should yield 2.5 mL of plasma. However, the amount of plasma will vary based on the hydration level of the individual from which the blood was drawn. Hence, a 5-mL draw of whole blood may give between 1.5 and 3 mL of plasma.

9. Normalization to external factors such as platelet count is often considered unnecessary in SELDI-TOF MS because SELDI-TOF MS utilizes a "differential display" and is often normalized to the total ion current. Some patients with severe disease or undergoing treatment with certain drugs can be thrombocytopenic (a condition marked by low platelet counts) making it difficult to compare these patients to a "normal" (healthy) population as differences may be due to the low number of platelets and, hence, be non-specific. We suggest that results from SELDI-TOF, or other standard biochemical assays, can be normalized to platelet count and expressed as a quantity of protein per 100,000 platelets. Although, as stated before,

normalization, other than to total ion current, may not be required with SELDI-TOF MS as normalization to platelet count does not appear to alter results by a significant amount (37). Alternatively to platelet count, other traditional normalization procedures used in Western blots, and other biochemical techniques, such as actin or GAPDH levels, may be employed as long as the platelet pellet is relatively free of white and red blood cells and the levels of the molecule used to normalize is reflective of the platelet count. If it has been determined that the platelet pellets are relatively free of other cell types, a cellular protein should be used for normalization to ensure that plasma proteins are not affecting the normalization procedure. No matter how much care is taken, a minute volume of plasma will be retained within the platelet pellet, thus total protein should not be used as a normalization factor. Plasma protein levels are ~50 mg/mL, therefore, if 20 µL of plasma is left on the platelet pellet, an increase of 1 mg of protein may be observed. This is a significant increase over the typical 1–1.5 mg of total protein typically obtained from the platelet pellet obtained from 1 mL of human PRP. It is our belief that platelet count is the optimal method for normalization in cases where external normalization is required.

10. The entire amount of mouse platelet lysate is used for analysis as collection from mouse to mouse can vary in the amount of PRP obtained, whereas in humans a consistent 1 mL of PRP is utilized for analysis. Often, only 200 µL of PRP can be obtained from a mouse. In order to ensure high quality spectra, the entire platelet lysate must be used.

References

1. Krueger, K.E. 2006. The potential of serum proteomics for detection of cancer: promise or only hope? *Onkologie.* **29**:498–499.
2. Huang, L.J., Chen, S.X., Huang, Y., Luo, W.J., Jiang, H.H., Hu, Q.H., Zhang, P.F., and Yi, H. 2006. Proteomics-based identification of secreted protein dihydrodiol dehydrogenase as a novel serum markers of non-small cell lung cancer. *Lung Cancer.* **54**:87–94.
3. Barker, P.E., Wagner, P.D., Stein, S.E., Bunk, D.M., Srivastava, S., and Omenn, G.S. 2006. Standards for plasma and serum proteomics in early cancer detection: a needs assessment report from the national institute of standards and technology--National Cancer Institute Standards, Methods, Assays, Reagents and Technologies Workshop, August 18–19, 2005. *Clin Chem.* **52**:1669–1674.
4. Kawada, N. 2006. Cancer serum proteomics in gastroenterology. *Gastroenterology.* **130**: 1917–1919.
5. Wu, G.H., Wang, Y.M., Yen, A.M., Wong, J.M., Lai, H.C., Warwick, J., and Chen, T.H. 2006. Cost-effectiveness analysis of colorectal cancer screening with stool DNA testing in intermediate-incidence countries. *BMC. Cancer.* **6**:136.
6. Lim, S.B., Jeong, S.Y., Kim, I.J., Kim, D.Y., Jung, K.H., Chang, H.J., Choi, H.S., Sohn, D.K., Kang, H.C., Shin, Y. et al 2006. Analysis of microsatellite instability in stool DNA of patients with colorectal cancer using denaturing high performance liquid chromatography. *World J Gastroenterol.* **12**:6689–6692.
7. Half, E.E., and Lynch, P.M. 2006. Mutated DNA in the stool--does it have a role in colorectal cancer screening? *Nat. Clin Pract. Gastroenterol. Hepatol.* **3**:594–595.
8. Zou, H., Harrington, J.J., Klatt, K.K., and Ahlquist, D.A. 2006. A sensitive method to quantify human long DNA in stool: relevance to colorectal cancer screening. *Cancer Epidemiol. Biomarkers Prev.* **15**:1115–1119.

9. Watanabe, T., Kobunai, T., Toda, E., Yamamoto, Y., Kanazawa, T., Kazama, Y., Tanaka, J., Tanaka, T., Konishi, T., Okayama, Y. et al 2006. Distal colorectal cancers with microsatellite instability (MSI) display distinct gene expression profiles that are different from proximal MSI cancers. *Cancer Res.* **66**:9804–9808.

10. Kreike, B., Halfwerk, H., Kristel, P., Glas, A., Peterse, H., Bartelink, H., and van, d., V 2006. Gene expression profiles of primary breast carcinomas from patients at high risk for local recurrence after breast-conserving therapy. *Clin Cancer Res.* **12**:5705–5712.

11. Chang, Y., and Liu, B. 2006. Difference of gene expression profiles between Barrett's esophagus and cardia intestinal metaplasia by gene chip. *J Huazhong. Univ Sci. Technolog. Med. Sci.* **26**:311–313.

12. Asgharzadeh, S., Pique-Regi, R., Sposto, R., Wang, H., Yang, Y., Shimada, H., Matthay, K., Buckley, J., Ortega, A., and Seeger, R.C. 2006. Prognostic significance of gene expression profiles of metastatic neuroblastomas lacking MYCN gene amplification. *J Natl. Cancer Inst.* **98**:1193–1203.

13. Folkman, J. 1971. Tumor angiogenesis: therapeutic implications. *N. Engl. J. Med.* **285**: 1182–1186.

14. Perez-Atayde, A.R., Sallan, S.E., Tedrow, U., Connors, S., Allred, E., and Folkman, J. 1997. Spectrum of tumor angiogenesis in the bone marrow of children with acute lymphoblastic leukemia. *Am. J. Pathol.* **150**:815–821.

15. Ribatti, D., Vacca, A., Nico, B., Quondamatteo, F., Ria, R., Minischetti, M., Marzullo, A., Herken, R., Roncali, L., and Dammacco, F. 1999. Bone marrow angiogenesis and mast cell density increase simultaneously with progression of human multiple myeloma. *Br. J. Cancer.* **79**:451–455.

16. Fuhrmann-Benzakein, E., Ma, M.N., Rubbia-Brandt, L., Mentha, G., Ruefenacht, D., Sappino, A.P., and Pepper, M.S. 2000. Elevated levels of angiogenic cytokines in the plasma of cancer patients. *Int. J. Cancer.* **85**:40–45.

17. Nguyen, M. 1997. Angiogenic factors as tumor markers. *Invest New Drugs.* **15**:29–37.

18. Dosquet, C., Coudert, M.C., Lepage, E., Cabane, J., and Richard, F. 1997. Are angiogenic factors, cytokines, and soluble adhesion molecules prognostic factors in patients with renal cell carcinoma? *Clin. Cancer Res.* **3**:2451–2458.

19. Abendstein, B., Daxenbichler, G., Windbichler, G., Zeimet, A.G., Geurts, A., Sweep, F., and Marth, C. 2000. Predictive value of uPA, PAI-1, HER-2 and VEGF in the serum of ovarian cancer patients. *Anticancer Res.* **20**:569–572.

20. Wong, A.K., Alfert, M., Castrillon, D.H., Shen, Q., Holash, J., Yancopoulos, G.D., and Chin, L. 2001. Excessive tumor-elaborated VEGF and its neutralization define a lethal paraneoplastic syndrome. *Proc. Natl. Acad. Sci. USA* **98**:7481–7486.

21. Rak, J., Klement, P., and Yu, J. 2006. Genetic determinants of cancer coagulopathy, angiogenesis and disease progression. *Vnitr. Lek.* **52 Suppl 1**:135–138.

22. Johnson, R.A., and Roodman, G.D. 1989. Hematologic manifestations of malignancy. *Dis. Mon.* **35**:721–768.

23. Langer, H., May, A.E., Daub, K., Heinzmann, U., Lang, P., Schumm, M., Vestweber, D., Massberg, S., Schonberger, T., Pfisterer, I. et al 2006. Adherent platelets recruit and induce differentiation of murine embryonic endothelial progenitor cells to mature endothelial cells in vitro. *Circ. Res.* **98**:e2–10.

24. Werther, K., Bulow, S., Hesselfeldt, P., Jespersen, N.F., Svendsen, M.N., and Nielsen, H.J. 2002. VEGF concentrations in tumour arteries and veins from patients with rectal cancer. *APMIS.* **110**:646–650.

25. Verheul, H.M., and Pinedo, H.M. 1998. Tumor Growth: A Putative Role for Platelets? *Oncologist.* **3**:II.

26. Verheul, H.M., Hoekman, K., Luykx-de Bakker, S., Eekman, C.A., Folman, C.C., Broxterman, H.J., and Pinedo, H.M. 1997. Platelet: transporter of vascular endothelial growth factor. *Clin. Cancer Res.* **3**:2187–2190.

27. Klement, G., Yip, T.-T., Cassiola, F., Kikuchi, L., Cervi, D., Podust, V.N., Italiano, J.E., Jr., Wheatley, E., Abou-Slaybi, A., Bender, E. et al 2009. Platelets actively sequester angiogenesis regulators. *Blood.* **113**:2835–2842.

28. Italiano, J., Richardson, J.L., Folkman, J., and Klement, G. 2006. Blood Platelets Organize Pro- and Anti-Angiogenic Factors into Separate, Distinct Alpha Granules: Implications for the Regulation of Angiogenesis. *ASH Annual Meeting Abstracts.* **108**:393.

29. Cervi, D., Yip, T.T., Bhattacharya, N., Podust, V.N., Peterson, J., bou-Slaybi, A., Naumov, G.N., Bender, E., Almog, N., Italiano, J.E.J. et al 2008. Platelet-associated PF-4 as a biomarker of early tumor growth. *Blood.* **111**:1201–1207.

30. Cervi, D., Yip, T.T., Bhattacharya, N., Podust, V.N., Peterson, J., bou-Slaybi, A., Naumov, G.N., Bender, E., Almog, N., Italiano, J.E.J. et al 2008. Platelet-associated PF-4 as a biomarker of early tumor growth. *Blood.* **111**:1201–1207.

31. Anderson, N.L., and Anderson, N.G. 2002. The human plasma proteome: history, character, and diagnostic prospects. *Mol Cell Proteomics.* **1**:845–867.

32. Davi, G., and Patrono, C. 2007. Platelet Activation and Atherothrombosis. *N Engl J Med.* **357**:2482–2494.
33. Michelson A.D. 2002. *Platelets.* Elsevier Science, Academic Press. San Diego, California, USA.
34. Hantgan, R.R., Taylor, R.G., and Lewis, J.C. 1985. Platelets interact with fibrin only after activation. *Blood.* **65**:1299–1311.
35. Addonizio, V.P., Jr., Fisher, C.A., Strauss, J.F., III, Wachtfogel, Y.T., Colman, R.W., and Josephson, M.E. 1986. Effects of verapamil and diltiazem on human platelet function. *Am J Physiol Heart Circ Physiol.* **250**:H366-H371.
36. 2004. *The Laboratory Mouse (Handbook of Experimental Animals).* Elsevier Academic Press. London.
37. Vermeulen, R., Lan, Q., Zhang, L., Gunn, L., McCarthy, D., Woodbury, R.L., McGuire, M., Podust, V.N., Li, G., Chatterjee, N. et al 2005. Decreased levels of CXC-chemokines in serum of benzene-exposed workers identified by array-based proteomics. *Proc. Natl. Acad. Sci. USA.* **102**:17041–17046.

Chapter 13

Using SELDI-TOF Mass Spectrometry on Amniotic Fluid and for Clinical Proteomics and Theranostics in Disorders of Pregnancy

Irina A. Buhimschi

Abstract

Clinical proteomics encompasses a multitude of experimental approaches, tools, and techniques based on proteomics technology which are directly aimed to accelerate and improve diagnosis and treatment of human diseases. Surface-enhanced laser desorption ionization time-of-flight (SELDI-TOF) mass spectrometry is a variant of matrix-enhanced laser desorption ionization (MALDI) that makes use of chemically-modified surfaces to reduce the complexity of biological samples prior to separation in the mass analyzer. Compared to other proteomic techniques, SELDI has several important advantages such as ability to analyze complex biological samples with minimal pre-processing, ease of handling and high throughput. Importantly, once the biomarker or combination of biomarkers with potential clinical value has been established, validation analyses can be conducted in close proximity to clinical settings which is important for establishing the utility of new diagnostics in clinical decision making and perhaps future theranostic interventions. This chapter provides protocols for experimental design and methodology aimed at (1) *discovering* biologically relevant biomarkers in amniotic fluid using SELDI-TOF; (2) *validating* the clinical utility of the biomarkers as new diagnostics; (3) *translating* the biomarker findings into pathophysiological phenomena to provide further insight and extend the current understanding of the disease process. Many of the principles described herein for amniotic fluid could be generalized to studies involving other types of biological samples and other clinical questions.

Key words: Biomarkers, Proteomics, SELDI, Mass spectrometry, Proteomics, Profile, Diagnostics

1. Introduction

1.1. SELDI-TOF as Diagnostics Platform for Amniotic Fluid Analysis

A biomarker is defined as a physical or biochemical characteristic that can be objectively assessed or measured as an indicator of "diseased" or "normal" state. The theory of medical diagnosis relies on the physicians' ability to recognize signs and symptoms of individual patients as biomarkers and corroborate them with their own prior experience or learned medical knowledge derived from the experience of their predecessors. The physician learns with practice to assign intuitively different certainty levels on each individual piece of information. In other words, when establishing a diagnosis some disease signs are weighed more heavily than others. However, after a comfortable level of certainty has been achieved, the physician will generally deduce a diagnosis and decide on treatment strategies based on the diagnosis. Difficulties arise because not all signs of disease are obviously detectable and therefore a certain degree of suspicion is required before deciding to perform a medical measurement test. Most diseases have a dynamic course and the skill is to corroborate a minimal number of signs and symptoms of early disease into a reasonable degree of suspicion to order a decisive diagnostic test. With respect to amniotic cavity, this results in performance of an amniocentesis procedure and retrieval of amniotic fluid which is then sent for laboratory analysis.

Depending on gestational age, the composition of amniotic fluid is the result of amniotic membrane secretion, fetal urine, and lung secretions. The fetus swallows amniotic fluid during most of pregnancy and thus participates directly to both the inflow and the outflow of substances in amniotic fluid. Therefore changes in fetal well-being are likely to result in earlier and more profound changes in composition of the amniotic fluid than of the maternal peripheral blood or urine. For this reason, the amniotic fluid is a preferred biological sample to start searching for proteomic biomarkers reflective of a fetal affection.

For amniotic fluid biomarker discovery, the principles of medical diagnosis described are the key in differentiating biomarkers of biological *relevance* (that should be weighed more upon in deciding final diagnosis) from *redundancy* (that could be discarded) and *noise* (that must be discarded). It is also important to understand that *biological relevance* does not overlap with *clinical utility* as the criteria for the latter depend on the disease that is studied, on the current state of clinical practice, and may change with time as more is learned about the disease and new drugs are discovered.

Within the last 50 years, advances in technology have facilitated measurements of various physical or biochemical characteristics and as a result the number of biologically relevant biomarkers has increased tremendously. Of these, however, only a handful has proven clinically useful. Therefore before embarking on proteomics experimentation to further identify proteins that are differentially expressed in a certain disease state, one needs to clearly set out the goals in terms of whether biomarkers are sought to mine further the disease process (biological relevance), to improve diagnosis in the current state of clinical practice (clinical relevance), or both.

Surface-enhanced laser desorption ionization time-of-flight (SELDI-TOF) mass spectrometry is a variant of matrix-enhanced laser desorption ionization (MALDI) that makes use of chemically-modified surfaces to reduce the complexity of biological samples prior to separation in the mass analyzer. SELDI has several important advantages such as the ability to analyze complex biological samples with minimal pre-processing, ease of handling, high throughput and requirement for small amounts of sample. Importantly, once the biomarker or combination of biomarkers with potential clinical value has been established, validation analyses can be conducted in close proximity to clinical settings, which is important to establish utility of new diagnostics for clinical decision making. Critics of SELDI-TOF often quote mass accuracy of the current technology as a limitation relative to other mass-spectrometry platforms as well as difficulty with biomarker identification for peaks of low intensity. Yet, an appropriate experimental design is generally able to overcome these alleged limitations. The ability to conduct experiments with larger numbers of clinical specimens thereby filtering through the biological variability is likely to result in more robust proteomic biomarkers for which identification will not be problematic.

1.2. General Principles of Experiment Design

Just as bedside medical diagnosis is both art and science, so is proteomics-based biomarker discovery and certain *choices* have to be made *a priori* from experimentation. These choices refer to: (1) the disease, (2) the biological sample, and (3) the proteomic techniques. In Table 1, we provide a few general guidelines which are derived from our experience as applied to disorders of pregnancy and to SELDI-TOF as the choice of initial proteomic technique, but these principles can be easily extrapolated to other clinical scenarios. In many instances, however, the choices are driven by combinations of less than ideal situations such as availability of a certain biological sample rather than its relevance for the disease of interest. As a result, the expectation for successful discovery of biomarkers with clinical utility should be lower. Unfortunately, many unsuccessful attempts in biomarker discovery are often blamed on the technology rather than the poor choices of experimental design. On the other hand, it is possible that the current clinical practice will benefit from even small improvements in diagnostics or that the newly discovered biomarkers will point to new means for therapeutical intervention, which could not have been envisioned by hypothesis-driven approaches. These possibilities certainly make some of the less than ideal experimental undertakings worthwhile, but also underscore the importance of an educated and carefully thought-out experimental design that can minimize disappointment and maximize the chance of success.

In our laboratory, we embarked on discovery of biomarkers with both biological and clinical relevance for prediction of preterm birth using proteomics, on the premise that alterations at protein levels may relate closer to the disease process than alterations in genes or mRNA. Our choice of using SELDI-TOF as

Table 1
Decisions to be made prior to experimentation

Choice	Question	Answer most likely to yield greatest success	Answer likely to yield lower success rates
Choice of disease	1. Is the disease important?	1. Disease is highly prevalent and/or outcome is unacceptable (severe disability or death)	1. Disease is rare and/or outcome is acceptable
	2. Is there a current need for better diagnostics?	2. Current diagnostics predict outcome poorly and/or have long turnaround time and/or a wrong diagnosis leads to treatment decisions with consequences that are final and cannot be reversed	2. No, several diagnostic test(s) are implemented in practice and/or are used to guide treatment and/or the adverse effects of treatment are minimal
	3. Is there a clear distinction between patients who have severe disease and those who do not?	3. Yes, but only in advanced stages where treatment options are limited	3. No, current diagnosis is clinical (subjective) and/or treatment interrupts disease progress before disease can be clearly distinguished
	4. Is there considerable patient heterogeneity?	4. None or minimal	4. Yes, the disease is likely a syndrome with multiple etiologies and pathogenic pathways converging in a single clinical scenario
	5. If diagnostic is perfected (higher accuracy, earlier detection) would it change clinical management?	5. Yes, doctors need to decide on diagnosis quickly and/or accurately and/or the decision impacts treatment options and outcome	5. No, there is no treatment available and/or adverse effects and cost of current treatment are minimal
Choice of biological sample	1. Is the sample relevant for the disease process?	1. Yes, the biological sample is in spatial and temporal proximity of an affected organ which is typically affected part of the disease	1. No, the biological sample is remote from affected organs, or the onset of disease is after the sample is available
	2. Is the sample likely to be changed in response to unrelated diseases in the same population?	2. No or likely in a different way	2. Yes, disease states of other systems or organs are likely to change the composition of the sample in a similar way

(continued)

Table 1 (continued)

Choice	Question	Answer most likely to yield greatest success	Answer likely to yield lower success rates
	3. Is the sample easily available?	3. Yes, sample can be obtained safely and non-invasively (i.e., urine, saliva) or minimally invasive (i.e., blood, tear fluid)	3. No, invasive surgical procedures are required to obtain the sample and/or specialized training (i.e., internal organ tissue or fluid)
	4. Is the sample available repeatedly?	4. Yes, sample can be obtained repeatedly without discomfort for the patient	4. No, invasive procedure cannot be repeated without substantial risk or discomfort for the patient
	5. Is the sample likely to degrade rapidly ex vivo?	5. No, sample is a fluid that accumulates in vivo inside a body cavity. Proteolysis likely has already occurred in vivo before sample is obtained	5. Yes. Sample is a tissue rich in proteases

proteomics platform was based on the reasoning that minimal sample handling, high throughput, and speed of analysis in a clinical setting were more important than mass accuracy or protein identification, especially for the discovery phase. The two diseases that have been at the center of our attention are spontaneous preterm delivery and preeclampsia, which are two important obstetrical syndromes, main determinants of preterm birth and of maternal and neonatal morbidity and mortality. Both diseases are especially difficult to tackle with respect to the development of new diagnostics due to patient heterogeneity (i.e. different etiologies and pathogenic pathways converging into identical clinical scenarios), difficulty in obtaining relevant biological samples from the fetus, and subjective gold standards for presence and absence of disease. Despite these difficulties, by using the experimental design described in the following section, we were able to extract several proteomic profiles with both biological and clinical significance. One amniotic fluid proteomic profile, the MR (Mass Restricted) score composed of four biomarker peaks, appears best to diagnose intra-amniotic inflammation leading to preterm delivery and predict histological chorioamnionitis, funisitis, and early onset neonatal sepsis (1–8). Another amniotic fluid proteomic profile composed of five different peaks was found to occur with increased frequency in idiopathic spontaneous preterm birth which by our definition encompasses women who do not exhibit the proteomic profile characteristic for intra-amniotic inflammation or intra-amniotic bleeding yet still deliver preterm (9).

In addition to amniotic fluid, we have applied the same principles to other biological sample such as cerebrospinal fluid and urine. One proteomic profile composed of four different peaks in cerebrospinal fluid identifies subclinical subarachnoidal bleeding such as that in the context of severe preeclampsia (PPB score) (10) while a urine proteomic profile (UPS score) composed of 13 biomarker peaks was demonstrated as having the highest accuracy to diagnose severe preeclampsia and predict an early indicated delivery (11). This chapter is centered on experimental details which we believe have been critical to our successful experimentation but have remained to the most part unpublished.

Each project performed in our laboratory can be dissected into three phases each with one goal and a number of steps as indicated in Table 2. The three phases do not necessarily need to be sequential.

Table 2
General scheme of biomarker discovery using SELDI-TOF in our laboratory

Experimental phases	Goal	Steps
Exploratory phase (i.e. biomarker discovery)	Extract "best" combination of SELDI biomarkers from a disease-relevant biological sample	*Step 1*: Optimization of SELDI conditions using wide SELDI profiling and pools of samples from "diseased" and "non-diseased" subjects *Step 2*: Mass Restriction (MR) analysis, definition of SELDI diagnostic profiles and scores *Step 3*: Accuracy calculation of SELDI scores in the exploratory phase cohort (individual samples)
Challenge phase (i.e. biomarker validation)	Determine clinical utility of proteomic scores as diagnostic and predictive tools in the context of current clinical practice	*Step 1*: Challenge phase in cross-sectional cohort and calculation of diagnostic accuracy of SELDI proteomic scores *Step 2*: Challenge phase in serial cohort and calculation of predictive ability of SELDI proteomic scores *Step 3*: Determination of reliability (intra- and inter-assay variations) and stability of proteomic scores
Translational phase (i.e. biomarker pathophysiological relevance)	Extend knowledge on the disease process and discover potential therapeutical targets for a future theranostics trial	*Step 1*: Biomarker identification *Step 2*: Hierarchical cohort clustering based on proteomic scores *Step 3*: Identification-centered proteomic techniques on samples selected based on SELDI proteomic scores *Step 4*: Hypothesis-driven experiments based on biomarker identity

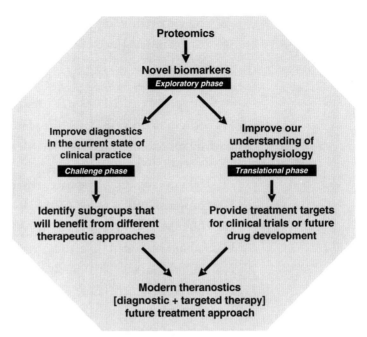

Fig. 1. Timeline of proteomics workflow in our laboratory. Republished from Am. J. Obstet. Gynecol. 2008 (12) with permission from Elsevier.

For efficiency purposes, we prefer they overlap in time. A critical issue, however, is that patient enrollment and biological specimen collection is progressing prospectively throughout the entire timeline of the study. The inclusion criteria relate to the physician's judgment to either perform a procedure or order tests to rule out the disease of interest in the real clinical setting. In our case, patients with signs and symptoms of spontaneous preterm birth where the physician has a suspicion for intra-amniotic infection are managed with amniocentesis and a battery of five tests is performed on amniotic fluid. These patients make up the *rule-out inflammation-induced preterm birth cohort*. Of note is that once the exploratory phase has been completed, the challenge and translational phases can progress in a parallel timeline for enhanced productivity as schematically represented in Fig. 1 (12).

2. Materials

2.1. Reagents

1. Centrifuge tubes (0.5, 1.5, 15, 50 ml).
2. 1.5 ml centrifuge dark colored tubes (for matrix dilutions).
3. PapPen (hydrophobic pen).
4. Acetonitrile (ACN), HPLC grade diluted 20 and 50% with ultrapure water.

5. Trifluoroacetic acid (TFA) (1 ml ampules, Pierce, Rockford, IL).
6. Protein assay (bicinchoninic acid/cupric sulfate (BCA) kit; Pierce).
7. Hydrochloric acid (HCl, 5 M, for pH adjustment).
8. Acetic acid (glacial, for pH adjustment).
9. Sodium hydroxide (10N, for pH adjustment).
10. 100 mM sodium acetate (NaOAc) adjusted to pH 4 with acetic acid.
11. 100 mM sodium phosphate buffer pH 7.
12. Phosphate buffer saline (PBS) (1×) pH 7.4.
13. 100 mM Tris–HCl, pH 8.
14. Sodium carbonate (Na_2CO_3) adjusted to pH 9.
15. 100 mM nickel (Ni^{2+}) sulphate aqueous solution.
16. 100 mM copper (Cu^{2+}) sulphate aqueous solution.
17. 100 mM zinc (Zn^{2+}) sulphate aqueous solution.
18. 100 mM gallium (Ga^{3+}) sulphate aqueous solution.
19. ProteinChip arrays: reverse phase hydrophobic surfaces (H4 and H50), anion/cation exchangers (Q10 and CM10) or immobilized metal affinity coupling (IMAC30) from BioRad Laboratories, Hercules, CA USA.
20. Matrix solvent: 50% (v/v) acetonitrile (ACN) and 0.5% (v/v) trifluoroacetic acid (TFA).
21. Energy-absorbing molecules (matrix): alpha-cyano-4-hydroxy cinnamic acid (CHCA), sinapinic acid (SPA) or EAM-1, 5 mg/tube (Bio-Rad) prepared as indicated in package insert to obtain a saturated matrix solution. Centrifuge briefly. Using the dark-colored tubes further dilute with matrix solvent to obtain a 20% saturated CHCA solution, 50% saturated SPA or 50% saturated EAM-1. Diluted matrix samples are stable up to 1 week.
22. Biological samples (frozen at −80°C and access to clinical source of fresh samples).

2.2. Equipment

1. Refrigerated centrifuge.
2. Deep freezer (−80°C).
3. Spectrophotometer (for protein assay per kit instructions).
4. Medium vacuum for spot-washing the arrays.
5. Pipettes and tips (low retention).
6. SELDI TOF ProteinChip reader (PBSIIC or PCS4000) externally calibrated using the $(M+H)+$ ion peaks of 1,296.5 Da (angiotensin), 1,570.6 Da ([Glu1] fibrinogen), 2,147.5 Da (porcine dynorphin A (209-225)) and 5,733.6 (bovine insulin) (BioRad SELDI mass standards).

3. Methods, Experiment Design and Data Analysis

3.1. Exploratory Phase

The exploratory phase may begin when biological samples (of choice) from approximately 20 cases in the "diseased" category and 20 cases in the "non-diseased" category have been gathered (see Note 1). The "diseased" cases have to be intentionally selected as having at least three clinical criteria of disease presence and/or severity including poor outcome. This method is detailed in particular for amniotic fluid but can be applied to other low complexity fluids such as cerebrospinal fluid, urine, peritoneal fluid, and tracheal aspirate. The examples are from experiments with either amniotic fluid or urine.

3.1.1. Step 1: Optimization of SELDI Conditions Using Wide SELDI Profiling and Pools of Samples from "Diseased" and "Non-diseased" Subjects

1. Measure total protein concentration and test if there is a statistically significant difference between the diseased and non-diseased groups. If yes, then protein normalization is strongly encouraged.

2. Prepare four composite pools (two diseased and two non-diseased) by mixing together different volumes of ten samples per pool (either diseased or non-diseased) so that each individual sample contributes to its pool with an equal amount of total protein.

3. Measure total protein of the four pools using the same assay as before and equalize/dilute protein concentration for all pools at 2 mg/ml with water (see Note 2).

4. Prepare the nine different types of binding/washing buffers listed in Table 3. For each energy-absorbing molecule, 16 arrays are required. CHCA works best for low molecular weight biomarkers (<10 kDa) while SPA is preferred for desorption of proteins >9 kDa. Generally CHCA and SPA are sufficient in this step. EAM-1 is a proprietary matrix formulation recommended for proteins of 10–50 kDa and for glycosylated proteins and could be tested to improve resolution of peaks which appear broad on the initial SPA SELDI tracings.

5. Prepare working dilutions (8-fold) of normalized pools in each of the respective binding/washing buffer. The working dilutions should have about 0.25 mg/ml total protein. Spot volumes of 6 μl (containing 1.5 μg total protein) will be applied per spot. An example of the array layout for this experiment is shown in Table 4.

6. Prepare the arrays for on-spot sample application. We prefer on-spot application and individual spot washings because of the small sample volumes required. Alternatively, a bioprocessor may be used and the array may be washed in-bulk. For our protocol, one-spot application volume refers to 6 μl/spot. One-wash volume refers to 120 μl partitioned into six applications (of about 20 μl each) which are rapidly applied on spot

Table 3
Optimized combination of array surfaces and binding/washing buffers for wide profiling

Arrays	Binding/washing buffer
H4	20% ACN
H4	50% ACN
H50	0.1% TFA
H50	15% ACN + 0.1% TFA
CM10	pH 4 (100 mM NaOAc)
CM10	pH 8 (100 mM Tris–HCl)
Q10	pH 6 (100 mM NaH_2PO_4)
Q10	pH 9 (100 mM Na_2CO_3)
IMAC30-metal	pH 4 (100 mM NaOAc)
IMAC30-metal	pH 7.4 (PBS)

Table 4
Array layout for wide profiling using pools of biological samples in Step 1 of Exploratory Phase

Spot	H4 arrays (CHCA and SPA)	H50 arrays (CHCA and SPA)	CM10 arrays (CHCA and SPA)	Q10 arrays (CHCA and SPA)
A	20% ACN Pool D1	0.1% TFA Pool D1	pH 4 Pool D1	pH 6 Pool D1
B	20% ACN Pool D1	0.1% TFA Pool D2	pH 4 Pool D2	pH 6 Pool D2
C	20% ACN Pool ND1	0.1% TFA Pool ND1	pH 4 Pool ND1	pH 6 Pool ND1
D	20% ACN Pool ND2	0.1% TFA Pool ND2	pH 4 Pool ND2	pH 6 Pool ND2
E	50% ACN Pool D1	15% ACN + 0.1% TFA Pool D1	pH 8 Pool D1	pH 9 Pool D1
F	50% ACN Pool D2	15% ACN + 0.1% TFA Pool D2	pH 8 Pool D2	pH 9 Pool D2
G	50% ACN Pool ND1	15% ACN + 0.1% TFA Pool ND1	pH 8 Pool ND1	pH 9 Pool ND1
H	50% ACN Pool ND2	15% ACN + 0.1% TFA Pool ND2	pH 8 Pool ND2	pH 9 Pool ND2

(continued)

Table 4 (continued)

Spot	IMAC-Zn^{2+} (CHCA and SPA)	IMAC-Cu^{2+} (CHCA and SPA)	IMAC-Ni^{2+} (CHCA and SPA)	IMAC-Ga^{3+} (CHCA and SPA)
Spot	IMAC-Zn^{2+} arrays (1: CHCA and 2: SPA)	IMAC-Cu^{2+} arrays (1: CHCA and 2: SPA)	IMAC-Ni^{2+} arrays (1: CHCA and 2: SPA)	IMAC-Ga^{3+} arrays (1: CHCA and 2: SPA)
A	pH 4 Pool D1	pH 4 Pool D1	pH 4 Pool D1	pH 4 Pool D1
B	pH 4 Pool D2	pH 4 Pool D2	pH 4 Pool D2	pH 4 Pool D2
C	pH 4 Pool ND1	pH 4 Pool ND1	pH 4 Pool ND1	pH 4 Pool ND1
D	pH 4 Pool ND2	pH 4 Pool ND2	pH 4 Pool ND2	pH 4 Pool ND2
E	pH 7.4 Pool D1	pH 7.4 Pool D1	pH 7.4 Pool D1	pH 7.4 Pool D1
F	pH 7.4 Pool D2	pH 7.4 Pool D2	pH 7.4 Pool D2	pH 7.4 Pool D2
G	pH 7.4 Pool ND1	pH 7.4 Pool ND1	pH 7.4 Pool ND1	pH 7.4 Pool ND1
H	pH 7.4 Pool ND2	pH 7.4 Pool ND2	pH 7.4 Pool ND2	pH 7.4 Pool ND2

D1 and *D2* two pools obtained from samples of diseased subjects
ND1 and *ND2* pools from samples of non-diseased subjects

and then aspirated to create a flow of wash buffer over the spot. It generally takes about 2–3 min to wash sequentially all eight spots of an array.

(a) H4 array preparation: circle each spot with PapPen and air dry. Precondition each spot by applying two 6 μl volumes of wash buffer and incubating for 5 min each.

(b) H50 array preconditioning: to each spot add 50% ACN for 5 min, wash with the respective buffer then leave the last spot volume on each spot for 5 min. Aspirate and apply samples to respective spots.

(c) CM10 and Q10 arrays are pre-equilibrated with two 5-min changes of the respective washing/binding buffer prior to sample application.

(d) IMAC30 array preconditioning: apply 6 μl of respective metal solution to each spot, wait for 15 min, aspirate, and apply a second 6 μl volume for another 15 min. Wash with respective wash buffer and then apply the samples.

7. Incubate the arrays in a moist chamber for 1 h at room temperature (20°C).

8. Wash each spot with one wash volume of binding/washing buffer. Air-dry the array for 10 min.

9. Apply matrix solution (one application of 1 μl 20% CHCA or two sequential applications of 1 μl 50% SPA, allowing for the

spots to dry completely between applications). Air dry the array for 10 min.

10. Read each array under three laser intensities/detector sensitivities combinations (low-laser, mid-laser and high-laser). The array reading setting should be determined empirically by carefully examining signal-to-noise ratios for each reading as well as baseline elevations. The array-reading protocols used in our laboratory are listed in Notes 3 and 4 and can be used as starting point when using a PBSIIC system.

11. Examine carefully each tracing and determine the optimal experimental combinations of binding buffer and array surfaces. First eliminate conditions where few peaks are observed. Examine the remaining conditions by comparing the tracings of the four pools within each condition with the auto-scale option disabled. Search for at least two conditions with highest signal-to-noise ratio which complement each other in resolved peaks. The best scenario is to find conditions that resolve peaks that are present in "diseased" state, while absent in "non-diseased" state. For example, the tracings we illustrate in Fig. 2 resulted from a wide profiling screen of one diseased pool *(D)* and one non-diseased *(ND)* pool under four different conditions represented by the combination of the array, binding buffer, matrix, and reading conditions. Note that both the H4 and H50 conditions show unique peaks in D1 tracing compared to ND1. These peaks do not resolve at the same mass indicating that they may represent different biomarkers and that these two conditions complement each other. By contrast, IMAC-Ga^{3+} resolved the same markers as the H4 condition albeit at a lower signal-to-noise ratio. The Q10 array condition was not able to resolve any peaks above the noise level for this sample and thus this condition was eliminated without hesitation.

3.1.2. Step 2: Data Collection, Spectrum Processing, and Peak Detection

1. Next, all samples that have been used to prepare the pools ($n=40$) are run in the two or three selected conditions established during Subheading 3.1.1. To estimate, there will be six arrays needed for each chosen condition. On each array, one spot is assigned as "blank" and is spotted with the respective binding buffer alone. The blank spots serve to determine the level of "background noise". It is critical in this step for samples to be arranged on arrays at random and not to assign them to spots based on categories ("diseased", "non-diseased" or buffer alone). This simple experimental layout can avoid much of the bias related to possible array inconsistencies or any unexpected change in instrument settings. Such issues have been at the center of several controversies that emerged in earlier stages of biomarkers discovery using SELDI (13).

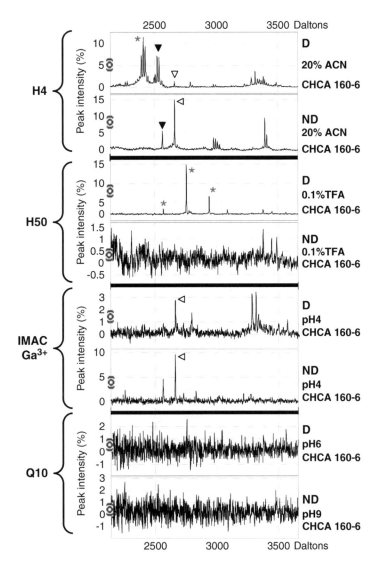

Fig. 2. Representative SELDI tracings obtained in Subheading 3.1.1 of the Exploratory Phase (wide profiling). *D*: pool of diseased samples (shown here: urine from women with severe preeclampsia). *ND*: pool of non-diseased samples (urine from normal pregnant women). CHCA: (alpha-cyano-4-hydroxy cinnamic acid). Other annotations refer to binding/washing conditions (ACN: acetonitrile, TFA: trifluoroacetic acid) and instrument reading settings such as laser intensity (160) and detector sensitivity (6). The *asterisks* note different biomarkers resolved on two complementing conditions (ideal situation and retained in final profile). The *solid black arrowheads* mark non-identical peaks, but disease marker resolves in close proximity to a peak present in non-diseased sample (eliminated based on Criterion V of the MR scoring algorithm (see Table 5)). The *open arrowheads* point to peaks representing the same peptide (i.e. resolved at the same *m/z*), which is not a disease marker (eliminated based on criterion III). After baseline subtraction, the tracings were normalized for total ion current (at 1,500 Da) as indicated by the *blue mark* (o) to the *left* of each tracing.

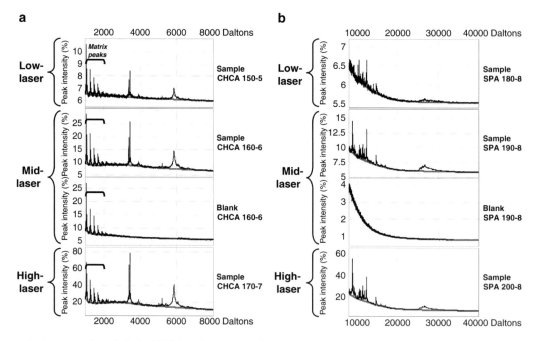

Fig. 3. Unprocessed amniotic fluid SELDI tracings obtained in Subheading 3.1.2 of the Exploratory Phase. These tracings are shown without baseline subtraction and without normalization. The baseline is marked by the *thick red line* and allows subjective judgment of reading conditions such as laser intensity and detector sensitivity which are the two variables we vary in our standardized array reading protocols. Panel (**a**) shows low molecular weight tracings of spots covered with alpha-cyano-4-hydroxy cinnamic acid (CHCA). Note that the matrix peaks also appear in the tracing from the blank spot. Panel (**b**) is a partial display of high molecular weight tracings from spots covered with sinapinic acid (SPA). The SPA matrix peaks are outside the displayed part of the spectrum. The instrument reading settings are listed in Notes 3 and 4.

2. Next, each array is read using three spot protocols with increasing level of laser intensity and/or detector sensitivity (see Notes 3 and 4 for reading settings used with our PBSIIC instrument). However, the exact reading settings will need to be determined empirically by critically examining the level of baseline elevation on tracings that received both samples and blank. As a guide, in Fig. 3, we display SELDI tracings (before baseline subtraction) obtained from one sample spot (amniotic fluid diseased pool: intra-amniotic inflammation). A tracing from the blank spots is shown for the mid-laser setting only. In the end, there will be about 135 tracings generated for each array/binding buffer combination chosen in Subheading 3.1.1. At least until now, we found that analyzing in this step just the tracings obtained with the mid-laser setting has been sufficient. Thus only from this reading setting we will have to analyze 40 SELDI tracings resulting from sample analyses and at least five blank tracings. The spectra obtained from the low- and high-laser reading settings are saved separately and used at the end of the exploratory phase to verify that the reading condition is

indeed optimal for the final discriminatory profile or to make minor adjustments to laser intensity and/or detector sensitivity before the challenge phase (see Subheading 3.1.4).

3. With all the tracings selected (using CRL-A command), subtract the baseline using the tool built into the ProteinChip software but do not use the normalization tool (η) or other filtering options in this step.

4. With all the tracings selected (CRL-A command), use the centroid peak tool built into the ProteinChip software to select all peaks that are visually detectable. Even when a peak is visually present in just one tracing, by selecting that peak in that tracing, the respective *m/z* value will be selected in all tracings given that the tool finds what it would automatically consider as peak even if practically no peptide was resolved at that mass (as in the case of the blank spots). The investigator selecting the peaks in this way will need to establish some rules for what he/she considers as peak. In our case, we would select a peak if it remains visible to the eye after covering the tracing above and below on the screen. Despite its tediousness, in this phase, we prefer this rather simple manual method to the automated tool of all peak detection.

5. Next, save the analyzed file separately and export to a Microsoft Excel spreadsheet (as comma separated value (CVS) file) the following peak information: Spectrum tag, spectrum #, peak #, intensity, (*m/z*) – 1 (substance mass), and signal-to-noise ratio (S/N).

3.1.3. Data Analysis

Mass Restriction Analysis, Definition of SELDI Diagnostic Profiles and Scores

Once the peak information has been obtained, you are ready to begin sorting out which peaks are informative and hence could classify as potential *biomarkers*. These have to be clearly filtered from peaks that carry *redundant* information that could be discarded, and peaks which belong to *noise* and thus must be discarded. In our laboratory, this procedure is performed by applying a sequential order of rules that together compose the algorithm of MR scoring. This is essentially a method of complexity reduction that avoids data over-fitting that has been optimized specifically for SELDI projects. Although there have been several automated software tools proposed for this purpose, none of these were able to achieve, at least in our hands, the performance of MR scoring. One of the reasons might be that the MR scoring method bases the analysis of proteomics data on a few practical principles of medical diagnosis rather than statistical conventions which, at least in our opinion, do not have yet the ability to fully encompass the complexity of human diseases. One of these is that by definition, "diseased" is a not a stand-alone state but relatively depends on what is considered "normal". In other words "one has to stand out from the crowd" to be classified as abnormal.

Table 5
Principles of the Mass Restricted (MR) scoring algorithm

Criterion	Principle
Criterion I	All biomarker peaks should be present in the "diseased" state (i.e., the search is for peaks that appear rather than disappear in disease state)
Criterion II	The final profile should include peaks from at least two different experimental conditions
Criterion III	All peaks in the final profile should be significantly different at a level of <0.0001 between the "diseased" and "normal" state (i.e. for each remaining biomarker there should be a highly significant difference in S/N ratios between diseased and non-diseased tracings)
Criterion IV (optional)	Only parent peaks should be considered (singly charged, least oxidized)
Criterion V	Peaks should not occur in areas where significant peaks appear in "normal" individuals as they could potentially interfere with informative peaks in "diseased" patients
Criterion VI (optional)	The final diagnostic proteomic profile should be minimal (restricted)

Republished from Am. J. Obstet. Gynecol. 2008 (11) with permission from Elsevier

1. To begin, each SELDI tracing is visually composed of sequences of "hypervariable" and "silent" areas or simply mass areas with peaks or just with noise, respectively. Mathematically, however, SELDI tracings are functions in a Euclidian plane defined by two Cartesian coordinates. By convention, we define a SELDI peak by three attributes: mass (peak mass = $(m/z) - 1$) which is the attribute of identity and the *x axis* coordinate, peak intensity (the attribute of relative abundance and the *y axis* coordinate) and peak presence (signal-to-noise ratio (S/N) greater than a cut-off of average S/N + 2 standard deviations (SD) for each corresponding mass value calculated from the tracings of the blank spots in the identical experimental condition including laser/detector settings). Peaks considered present are assigned a Boolean indicator of 1 as opposed to peaks below the cut-off which are assigned a Boolean indicator of 0.

2. The principles of MR scoring have been published earlier (4, 7, 11) and are listed together in Table 5. Next, the exported peak attributes are arranged in a spreadsheet and a Macro tool assigned to perform the necessary logical operations of the MR scoring algorithm. A representative section of a spreadsheet is shown in Fig. 4 which displays criteria I–III applied to three

		14.5 kDa	14.5 kDa	14.5 kDa	14.5 kDa	14.5 kDa	15.1 kDa	15.1 kDa	15.1 kDa	15.1 kDa	15.1 kDa	15.3 kDa	15.3 kDa	15.3 kDa	15.3 kDa	15.3 kDa
Code	Type	Mass	Real mass	Intensity	S/N	present?	Mass	Real mass	Intensity	S/N	present?	Mass	Real mass	Intensity	S/N	present?
Sample1	ND	14554.25	14554.25	0.13	3.91	1	15138.69		1.19	0.04	0	15306.50		0.04	1.47	0
Sample2	ND	14536.54	14536.54	0.20	6.08	1	15078.13		1.29	0.04	0	15343.02		0.04	1.24	0
Sample3	ND	14542.37	14542.37	0.35	10.85	1	15124.95		1.82	0.06	0	15283.11		0.05	1.63	0
Sample4	D	14542.97	14542.97	0.40	11.17	1	15120.37	15120.37	4.77	0.16	1	15334.57	15334.57	0.08	2.49	1
Sample5	ND	14543.09	14543.09	0.30	8.59	1	15078.13		1.44	0.05	0	15320.67		0.04	1.42	0
Sample6	D	14538.49	14538.49	0.37	10.01	1	15096.62	15096.62	3.94	0.14	1	15325.59		0.04	1.29	0
Sample7	D	14530.58	14530.58	0.35	9.10	1	15110.94	15110.94	23.90	0.84	1	15312.88	15312.88	0.28	8.28	1
Sample8	D	14540.84	14540.84	0.37	10.25	1	15097.56	15097.56	15.33	0.52	1	15310.94	15310.94	0.13	4.09	1
Sample9	ND	14545.59	14545.59	0.38	10.29	1	15166.98		1.96	0.07	0	15304.28		0.06	1.70	0
Sample10	ND	14547.46	14547.46	0.35	10.73	1	15157.92		2.21	0.07	0	15352.79		0.05	1.58	0
Sample11	ND	14546.59	14546.59	0.28	8.00	1	15147.30	15147.30	3.11	0.10	1	15358.60		0.04	1.34	0
Sample12	ND	14552.85	14552.85	0.52	12.59	1	15085.38	15085.38	1.83	0.07	0	15293.10		0.06	1.68	0
Sample13	D	14550.15	14550.15	0.59	15.05	1	15110.72	15110.72	11.54	0.42	1	15317.45	15317.45	0.12	3.28	1
Sample14	ND	14546.46	14546.46	0.24	7.63	1	15084.03		1.49	0.04	0	15267.36		0.05	1.63	0
Sample15	ND	14548.01	14548.01	0.30	9.75	1	15176.49		1.41	0.04	0	15290.58		0.05	1.60	0
Sample16	ND	14544.50	14544.50	0.46	12.91	1	15111.42	15111.42	2.26	0.08	1	15281.73		0.06	1.97	0
Sample17	ND	14548.07	14548.07	0.20	6.17	1	15106.36		1.53	0.05	0	15303.13		0.04	1.20	0
Sample18	D	14550.02	14550.02	0.20	6.43	1	15110.62	15110.62	5.88	0.17	1	15316.17	15316.17	0.07	2.40	1
Sample19	ND	14557.22	14557.22	0.14	5.15	1	15099.40		1.32	0.03	0	15299.18		0.02	0.82	0
Sample20	ND	14554.55	14554.55	0.22	7.52	1	15157.85		1.67	0.05	0	15356.54	15356.54	0.05	1.77	1
Sample21	ND	14548.85	14548.85	0.20	6.64	1	15081.09		1.01	0.03	0	15309.73		0.03	1.29	0
Sample22	ND	14544.76	14544.76	0.15	5.18	1	15147.39		1.41	0.04	0	15318.74		0.03	1.05	0
Sample23	ND	14544.06	14544.06	0.16	5.98	1	15135.73		1.41	0.04	0	15294.12		0.03	1.18	0
Blank		14535.72		0.03	1.35		15154.29		2.90	0.06		15344.02		0.03	1.38	
Blank		14552.00		0.05	1.64		15174.10		1.93	0.05		15333.48		0.06	1.72	
Blank		14514.06		0.02	0.89		15086.66		2.19	0.05		15330.15		0.07	1.76	
average S/N Blank					1.30					0.05					1.62	
SD S/N Blank					0.38					0.01					0.21	
dynamic cut-off (average + 2SD)					2.05					0.07					2.04	
average real mass		14546.01		0.30	8.69		15110.10		4.08	0.14		15324.76		0.06	2.02	
SD real mass		6.09		0.12	2.84		17.47		5.57	0.20		17.69		0.05	1.55	
number tracings		23.00		23.00	23.00		17.00		23.00	23.00				23.00	23.00	
number tracings with peak present						23					9					6
average Diseased				0.38	10.34				10.89	0.37				0.12	3.64	
average Nondiseased				0.27	8.12				1.67	0.05				0.04	1.45	
F-test				0.67641	0.79123				0.00000	0.00000				0.00000	0.00000	
P value < 0.0001				no	no				yes	yes				yes	yes	

Fig. 4. Representative spreadsheet calculations for the Mass Restricted (MR) scoring method. The MR method is an algorithm to extract peaks of biological relevance from SELDI tracings and consists of a set of rules which are sequentially applied to keep or eliminate peaks from the final profile.

SELDI peaks. Some of the criteria of the MR scoring method are disease- and sample-dependent and should be viewed as optional. For instance, it is possible that various degrees of oxidation will produce valuable biomarkers which may relate to the disease process (11) (criterion IV). Alternatively, oxidation may be an ex vivo phenomenon and thus will appear inconsistently among the diseased tracings and thus one should be cautious when selecting both the parent and the oxidized form into the final profile. Criterion V assures that the final biomarkers are absent in "non-diseased" state. This is of utmost importance because the final diagnostic profile will not rely merely on quantitative differences in intensity of peaks which could be impacted nonspecifically, for instance, by small variations in concentration but rather on generation of new peptides. In the rare instance where there are no unique peaks characteristic of the "diseased" state, Criterion III will still assure a highly significant difference in S/N between "diseased" and "non-diseased" states. The final criterion (VI, optional) refers to restriction or parsimony. In other words, the profile should

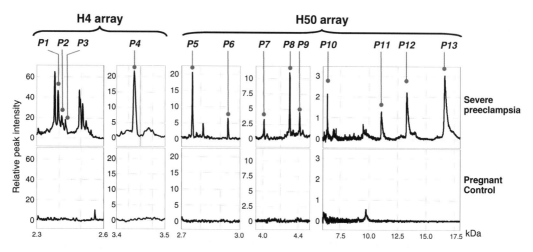

Fig. 5. Biomarker components of the urine proteomic score (UPS) characteristic of severe preeclampsia in need of indicated early delivery. Representative SELDI profile from a patient with severe preeclampsia at 30 weeks of pregnancy illustrating the 13 biomarker components (P1–P13, marked with *bullets*) of the urinary proteomic scores (Boolean UPSb = 13 and Ranked UPSr = 47) presented in comparison with a SELDI profile of urine from a healthy pregnant control patient (both UPSb and UPSr of 0) at the same gestational age. kDa: kilodaltons. Republished from Am. J. Obstet. Gynecol. 2008 (11) with permission from Elsevier.

contain the minimal number of peaks required for maximum discrimination. Applying this criterion is dependent on the clinical context where improving the diagnosis is sought. For instance, if the biological sample is the result of an invasive procedure, unlikely clinically to be repeated and a rapid result turnaround is necessary then restriction should be performed. This was the case both with the MR profile in amniotic fluid obtained at amniocentesis and the PPB profile in cerebrospinal fluid obtained by spinal tap. In contrast, if the purpose is to detect asymptomatic disease, the biological sample can be obtained non-invasively and thus the test can be easily repeated then the last criterion should not be applied. The reason for this is that a restricted profile will likely perform well in challenge phase at detecting severe and likely symptomatic disease but will not comprise a sufficient spectrum of biomarkers to perform adequately in milder or asymptomatic disease or in the presence of other co-morbidities. In our experience, in all clinical scenarios where we sought and then validated proteomic profiles we noted a sequential and ordered appearance of the biomarkers which ultimately comprised the final profile. The number of biomarkers and their intensity was in direct correlation with disease severity. Therefore, to be able to detect a proteomic abnormality in asymptomatic patients, one needs a large enough number of peaks to explore during the challenge phase. This was the case with the urine UPS profile of severe preeclampsia and the reason for the 13 markers required (Fig. 5).

3. After the final composition of unique biomarkers (defined by unique average masses) that fulfil the rules of the Mass Restricted algorithm is determined, this proteomic information is transformed into a named operator (score) which characterizes each sample and can be manipulated with regular statistics. This can be done in two ways. The simplest way is to sum the numbers of biomarkers of the final profile present in that particular sample. We think this applies best to restricted profiles such as the amniotic fluid MR score and cerebrospinal fluid PPB score. Thus, these scores ignore any quantitative information related to peak intensity and rely solely on the presence of peaks that are otherwise absent in the respective biological compartment. This is specifically advantageous when a rapid diagnosis is sought as normalization for total protein is not required to achieve the same accuracy. Both the MR score and the PPB score can return a test result within 15 min and do not require a standard curve (1, 10). For profiles obtained after protein normalization, however, adding the semiquantitative information in the equation will result in a continuous rather than ordinal result. Because continuous tests have higher power it is expected this type of operator will perform better in the challenge phase (14). This was exactly the case of the UPS score where the ranked variant (UPSr, continuous operator ranging from 0 to infinity) displaced the Boolean variant (UPSb, ordinal operator ranging from 0 to 13) when they were both entered into a multivariate model in the challenge phase (11).

3.1.4. Step 3: Verification of Accuracy in the Exploratory Phase Cohort

1. This step simply checks the accuracy of the new operators (SELDI scores) in segregating back the 40 samples used for their development based on the initial clinical criteria. In receiver operator curve analysis, using clinical diagnosis as gold standard, 100% sensitivity and 100% specificity should be expected since you are fitting back the model to the samples used to create it.

2. If outliers are observed we found it useful to review the medical records of those cases carefully before embarking onto the challenge phase. Because in the exploratory phase the diseased cases have been intentionally selected as having at least three disease criteria, outcome outliers are not expected unless the possibility for misdiagnosis in the current clinical practice is high. During the exploratory phases of the three projects to which we refer throughout the article, the only instance of misclassification in the exploratory phase was a case that lacked the cerebrospinal fluid profile characteristic of severe preeclampsia, yet the patient was clinically categorized as having severe preeclampsia by pulmonary edema in the absence of increased blood pressure or proteinuria. Our post-proteomic review of the medical record revealed that, according to the

obstetric care provider, the pulmonary edema likely resulted from iatrogenic fluid overload (10).

3. Accuracy calculations restricted to the exploratory phase sample will indicate a cut-off of maximum accuracy for this population. However, expect that this cut-off will likely need adjustment in the challenge phase given that the exploratory phase purposely omits cases with asymptomatic or mild disease.

4. These arrays are read with all three reading protocols (low-laser, mid-laser and high-laser) and scores are compared between the settings. The reading protocol displaying the largest level of significance between diseased and non-diseased samples will be used for the challenge phase.

3.2. Challenge Phase

3.2.1. Step 1: Challenge Phase in Cross-Sectional Cohort and Calculation of Diagnostic Accuracy of SELDI Proteomic Scores

1. The key for a successful challenge phase is to test the SELDI scores in a setting that is as close as possible to the one where the new diagnostics are most needed. Thus, in our setting, as soon as the SELDI method and MR score model have been finalized, we begin the challenge phase by testing virtually every case that is enrolled in our prospective cohort and has a clinical work-up for either preterm birth or preeclampsia ("intent to diagnose").

2. To increase the relevance of the challenge phase and to be able to judge the clinical utility of the SELDI scores, we perform the SELDI testing from fresh sample collection and generate the scores in blinded fashion in parallel with the clinical laboratory's testing and before a clinical diagnosis is confirmed or ruled out. In this phase, we also test samples from patients often encountered in the same clinical setting with related or unrelated diagnoses to examine the specificity of the SELDI profile for our disease of interest.

3. Because sample testing in the challenge phase is done based on availability, there will not always be eight patient samples to test at a time. We use an improvised modality to incubate samples on just few spots on the array as needed for each test while simultaneously keeping any unused spots dry. We also validated that the unused spots retain their surface-enhanced binding ability. In contrast, exposing unused spots to a humidifying environment diminish their reactivity and subsequently, cannot be used without a loss in surface affinity. In Fig. 6, we illustrate how this improvised and reusable device allows testing of fresh patient samples individually and economically during the challenge phase. Every reading is compared to a positive control tracing (with all biomarkers present) to indicate the m/z location of each of the sought biomarkers. Thereafter the final proteomic scores are calculated from the S/N ratios at each of the m/z values of interest.

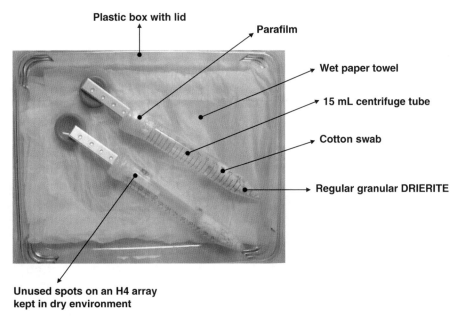

Fig. 6. Economical modality to use partial SELDI arrays in the Challenge Phase. The tube opening is covered with Parafilm. A central slit is cut into the Parafilm and the array carefully inserted inside the tube. After incubation, the blue cap is placed on the tube over the parafilm and the entire device reused in the next experiment.

4. If possible, the gold standard in the challenge phase should be independent of clinical diagnosis. In cancer studies, histological classification offers an acceptable gold standard. In contrast, in preterm birth or preeclampsia, the selection of an objective classification is hindered due to the fact that a diagnosis of disease will require the physician to interrupt the course of pregnancy per management protocols. However, by using an outcome variable (neonatal sepsis, histology of the placenta or indicated early delivery), the accuracy of SELDI scores can be compared using receiver operating curve (ROC) analysis with that of other laboratory tests accepted in clinical practice or research and development for the same purpose.

5. The number of cases needed for the challenge phase is a statistical matter and requires comment. Although there has been no statistical publication specifically targeting sample size calculation for proteomics operators derived from SELDI experimentation, the belief remains that a sample size calculation prior to initiating the challenge phase is absolutely necessary especially if clinical utility relative to other diagnostic tools is sought. Several software routines can be used for sample size estimation of diagnostic accuracy based on the algorithms of Hanley and Mc Neil (15) and/or Obuchowski (16). The PASS program by NCSS (Kayseville, UT; http://www.ncss.com)

and Power and Precision by Biostat (Englewood NJ; http://www.power-analysis.com) run on Microsoft Windows are both user friendly and can be evaluated by downloading free trial versions over the internet. To obtain a general estimate of how many cases are needed, the prevalence of the sought outcome in the challenge cohort needs to be known. For example, at our institution the prevalence of an indicated delivery for preeclampsia in cases with expressed intent to diagnose is about 40%. The accuracy of a clinically used test alternative (protein-to-creatinine ratio, measured as area under the curve) is less than 0.6 and that of a proposed alternative in research testing (ratio of urine angiogenic factors) is about 0.8. In order to detect a significant difference in area under the curve of the SELDI UPSr score versus ratio of angiogenic factors at a power level of 0.9 and an alpha of 0.5, we determined by using PASS 2008 software that a minimum of 66 cases are needed (25 positive and 41 negative for outcome). We decided, however, to stop the challenge phase only after enrolling at least double this calculated number on the empirical basis that we were expecting a high rate of misclassification, as currently there is no clinical means of discriminating severe preeclampsia from other proteinuric or hypertensive conditions complicating pregnancy. In the challenge phase, such cases were all targeted for enrollment. In Fig. 7, we present the distribution of proteomic scores UPSb and UPSr in the cases analyzed as part of the exploratory versus the challenge phase. Of note, although in the exploratory phase there is no overlap of scores between diseased (preeclamptic with indicated early delivery) and non-diseased cases (non-preeclamptic with normal delivery at term), a significant overlap is noted in the challenge phase despite all graphs statistically differing at $P<0.0001$. Yet, the fact that the proteomic score performs in the same population significantly better ($P<0.001$) than the other tests used or proposed for the same purpose indicates that the limitation lies not in our test but in the available gold standard. However, given its highly significant clinical relevance these results underscore the importance of providing clinicians with additional markers to aid with clinical decisions with significant impact. Urine SELDI scoring can find thus immediate application in current clinical practice by selecting the cases that need immediate intervention from those that can be management expectantly at least until repeat analysis.

3.2.2. Step 2: Challenge Phase in Longitudinal (Serial) Cohort

This step is important to perform when the goal is to test accuracy in detecting asymptomatic disease. As detailed earlier, an outcome variable should be used as gold standard. Otherwise, a test that truly detects subclinical disease will appear falsely positive if the

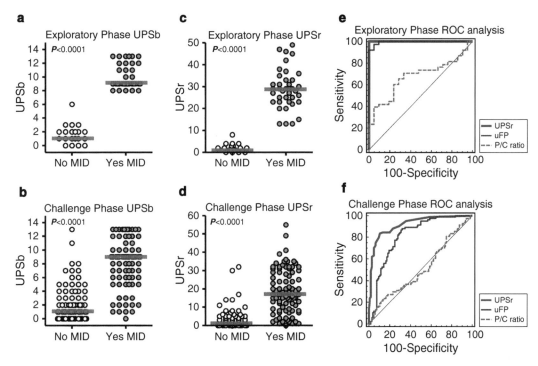

Fig. 7. Clustering of cases and receiver operating curve (ROC) analysis in the Exploratory versus Challenge Phase. Cases in the Exploratory (**a** and **c**) and Challenge Phases (**b** and **d**) are grouped by their urine proteomic scores (UPSb (ordinal variable) and UPSr (continuous variable)) and by the occurrence of poor outcome (MID: medically indicated delivery for preeclampsia). The *thick red line* in Panels **a–d** marks the group median. The figure illustrates that while there is no overlap between the two groups in the Exploratory Phase, we expect a significant overlap in the Challenge Phase where enrollment is consecutive based on "decision to diagnose" and the outcome is not natural but dictated by the physician's decision to intervene and interrupt the course of pregnancy. Despite the overlap, in comparative receiver operating (ROC) analyses, the SELDI score (*thick red line*) is superior to any other equivalent tests from the same sample such as protein-to-creatinine ration (P/C ratio, *green dotted line*) or ratio in angiogenic factors (uFP, *blue solid line*) (**e**: Exploratory Phase vs. **f**: Challenge Phase). Panel **f** is republished from Am. J. Obstet. Gynecol. 2008 (11) with permission from Elsevier.

clinical diagnosis at the time of sample collection is used for classification.

3.2.3. Step 3: Reliability and Stability of Biomarkers

When analyzing reliability (inter- and intra-rater variability) as well as stability of biomarkers, it is important to use agreement and not correlation analysis. Statistical algorithms such as Bland and Altman agreement and kappa coefficients are especially useful in evaluating robustness of the newly developed test (11). For inter and intra-rater variability, eight samples are tested in triplicate by three independent observers from preparation of the arrays to score calculation. For biomarker stability, scores from fresh samples are compared to those obtained after samples have been stored at either −80°C, −20°C, 4°C, or at room temperature for different periods of time (11).

3.3. Translational Phase

The translational phase requires identification of the proteins responsible for the peaks included in the final profile. In all instances where we employed the above described algorithm of discovery and validation, the identity of the proteins was highly informative and found to closely relate to the pathogenic process of the disease. The methods for identification of SELDI markers are beyond the focus of this chapter. Importantly, however, in all our studies the identification of the markers led us to pathogenic pathways that could not have been envisioned or detected solely through hypothesis-driven approaches.

4. Notes

1. This starting number of exploratory phase samples is ideal but practically it cannot always be achieved. In these cases, we prefer to start with smaller pools of severe cases rather than dilute the diseased group with milder forms.
2. This protein concentration is ideal but it cannot always be achieved as some biological samples are inherently more diluted in vivo (e.g. cerebrospinal fluid). In such cases, adjust the concentration of the pools so that by diluting 3-fold with binding/washing buffers, the concentration on the spot will remain about 0.2 mg/ml.
3. Standardized spot protocols used in our laboratory for low mass readings (CHCA spots).

   (a) Low-Laser protocol (LL: 150-5)

Set high mass to 20,000 Da, optimized from 2,000 to 5,000 Da
Set starting laser intensity to 150
Set starting detector sensitivity to 5
Focus by optimization center
Set Mass Deflector to 800 Da
Set data acquisition method to Seldi Quantitation
Set Seldi acquisition parameters 22. delta to 5. transients per to 15 ending position to 82
Set warming positions with 2 shots at intensity 155 and do not include warming shots
Process sample

(b) Mid-Laser protocol (ML: 160-6)

Set high mass to 20,000 Da, optimized from 2,000 to 5,000 Da
Set starting laser intensity to 160
Set starting detector sensitivity to 6
Focus by optimization center
Set Mass Deflector to 800 Da
Set data acquisition method to Seldi Quantitation
Set Seldi acquisition parameters 20. delta to 5. transients per to 15 ending position to 80
Set warming positions with 2 shots at intensity 165 and Don't include warming shots
Process sample

(c) High-Laser protocol (HL: 170-7)

Set high mass to 20,000 Da, optimized from 5,000 to 8,000 Da
Set starting laser intensity to 170
Set starting detector sensitivity to 7
Focus by optimization center
Set Mass Deflector to 800 Da
Set data acquisition method to Seldi Quantitation
Set Seldi acquisition parameters 24. delta to 5. transients per to 15 ending position to 84
Set warming positions with 2 shots at intensity 175 and do not include warming shots
Process sample

4. Standardized spot protocols for high mass readings (SPA or EAM covered spots).

(a) Low Laser protocol (LL: 190-8)

Set high mass to 100,000 Da, optimized from 8,000 to 40,000 Da
Set starting laser intensity to 190
Set starting detector sensitivity to 8
Focus by optimization center
Set Mass Deflector to 500 Da
Set data acquisition method to Seldi Quantitation

Set Seldi acquisition parameters 20. delta to 5. transients per to 15 ending position to 80
Set warming positions with 2 shots at intensity 195 and do not include warming shots
Process sample

(b) Mid Laser protocol (ML: 200-8)

Set high mass to 100,000 Da, optimized from 10,000 to 40,000 Da
Set starting laser intensity to 200
Set starting detector sensitivity to 8
Focus by optimization center
Set Mass Deflector to 500 Da
Set data acquisition method to Seldi Quantitation
Set Seldi acquisition parameters 22. delta to 5. transients per to 15 ending position to 82
Set warming positions with 2 shots at intensity 205 and do not include warming shots
Process sample

(c) High Laser protocol (HL: 210-10)

Set high mass to 100,000 Da, optimized from 10,000 to 40,000 Da
Set starting laser intensity to 210
Set starting detector sensitivity to 10
Focus by optimization center
Set Mass Deflector to 500 Da
Set data acquisition method to Seldi Quantitation
Set Seldi acquisition parameters 24. delta to 5. transients per to 15 ending position to 84
Set warming positions with 2 shots at intensity 215 and do not include warming shots
Process sample

Acknowledgments

The conditions for wide profiling have been largely optimized in collaboration with Dr. Rob Christner, PhD to whom the author is deeply indebted. This work was supported in part from National

Institute of Heath (NIH)/Eunice Kennedy Shriver National Institute of Child Health and Human Development (NICHD) grant RO1 HD47321 (I.A.B.).

References

1. Buhimschi, C.S., Bhandari, V., Hamar, B.D., Bahtiyar, M.O., Zhao, G., Sfakianaki, A.K., Pettker, C.M., Magloire, L., Funai, E., Norwitz, E.R., Paidas, M., Copel, J.A., Weiner, C.P., Lockwood, C.J. and Buhimschi, I.A. (2007). Proteomic profiling of the amniotic fluid to detect inflammation, infection, and neonatal sepsis. *PLoS Med.* **4**: e18.
2. Buhimschi, I.A., Christner, R. and Buhimschi, C.S. (2005). Proteomic biomarker analysis of amniotic fluid for identification of intra-amniotic inflammation. *BJOG.* **112**: 173–181.
3. Buhimschi, I.A., Buhimschi, C.S., Weiner, C.P., Kimura, T., Hamar, B.D., Sfakianaki, A.K., Norwitz, E.R., Funai, E.F. and Ratner, E. (2005). Proteomic but not enzyme-linked immunosorbent assay technology detects amniotic fluid monomeric calgranulins from their complexed calprotectin form. *Clin. Diagn. Lab. Immunol.* **12**: 837–844.
4. Buhimschi, I.A., Buhimschi, C.S., Christner, R. and Weiner, C.P. (2005). Proteomics technology for the accurate diagnosis of inflammation in twin pregnancies. *BJOG.* **112**: 250–255.
5. Buhimschi, C.S., Pettker, C.M., Magloire, L.K., Martin, R., Norwitz, E., Funai, E. and Buhimschi, I.A. (2005). Proteomic technology and delayed interval delivery in multiple pregnancies. *Int. J. Gynaecol. Obstet.* **90**: 48–50.
6. Weiner, C.P., Lee, K.Y., Christner, R., Buhimschi, I.A. and Buhimschi, C.S. (2005). Proteomic biomarkers that predict the clinical success of rescue cerclage. *Am. J. Obstet. Gynecol.* **192**: 710–718.
7. Buhimschi, I.A., Zambrano, E., Pettker, C.M., Bahtiyar, M.O., Paidas, M., Rosenberg, V.A., Thung, S., Salafia, C.M. and Buhimschi, C.S. (2008). Using proteomic analysis of the human amniotic fluid to identify histologic chorioamnionitis. *Obstet. Gynecol.* **111**: 403–412.
8. Buhimschi, C.S., Buhimschi IA, Abdel-Razeq S. Rosenberg, V.A., Thung, S.F., Zhao, G., Wang, E. and Bhandari, V. (2007). Proteomic biomarkers of intra-amniotic inflammation: relationship with funisitis and early-onset sepsis in the premature neonate. *Pediatr. Res.* **61**: 318–324.
9. Buhimschi, I.A., Zhao, G., Abdel-Razeq, S., Rosenberg, V.A., Thung, S.F. and Buhimschi, C.S. (2008). Multidimensional proteomics analysis of amniotic fluid to provide insight into the mechanisms of idiopathic preterm birth. *PLoS ONE.* **3**: e2049.
10. Norwitz, E.R., Tsen, L.C., Park, J.S., Fitzpatrick, P.A., Dorfman, D.M., Saade, G.R. Buhimschi, C.S. and Buhimschi, I.A. (2005). Discriminatory proteomic biomarkers analysis identifies free hemoglobin in cerebrospinal fluid of women with severe preeclampsia. *Am J Obstet Gynecol.* **193**: 957–64.
11. Buhimschi, I.A., Zhao, G., Funai, E.R., Harris, N., Sasson, I., Bernstein, I.M., Saade, G.R. and Buhimschi, C.S. (2008). Specific urinary fragments of SERPINA-1 and ALBUMIN are proteomic biomarkers of preeclampsia. *Am. J. Obstet. Gynecol.* **199**: 551.e1–551.e16.
12. Buhimschi, I.A. and Buhimschi, C.S. (2008). Proteomics of the amniotic fluid in assessment of the placenta. Relevance for preterm birth. *Placenta.* **29**: Suppl A: S95–101.
13. Hu, J., Coombes, K.R., Morris, J.S. and Baggerly, K.A. (2005). The importance of experimental design in proteomic mass spectrometry experiments: some cautionary tales. *Brief Funct. Genomic. Proteomic.* **3**: 322–331.
14. Mandrekar, J.N. and Mandrekar, S.J. (2005). Statistical methods in diagnostic medicine using SAS® software. *Proceedings of the 30th SAS Users group International Conference (SUGI).* Paper 211–230.
15. Hanley, J.A. and McNeil, B.J. (1983). A method of comparing the areas under receiver operating characteristic curves derived from the same cases. *Radiology.* **148**: 839–843.
16. Obuchowski, N. and McClish, D. (1997). Sample size determination for diagnostic accuracy studies involving binormal ROC curve indices. *Stat. Med.* **16**: 1529–1542.

Chapter 14

High Throughput Profiling of Serum Phosphoproteins/Peptides Using the SELDI-TOF-MS Platform

Lin Ji, Gitanjali Jayachandran, and Jack A. Roth

Abstract

Protein phosphorylation is a dynamic post-translational modification that plays a critical role in the regulation of a wide spectrum of biological events and cellular functions including signal transduction, gene expression, cell proliferation, and apoptosis. Determination of the sites and magnitudes of protein phosphorylation has been an essential step in the analysis of the control of many biological systems. A high throughput analysis of phosphorylation of proteins would provide a simple, logical, and useful tool for a functional dissection and prediction of biological functions and signaling pathways in association with these important molecular events. We have developed a functional proteomics technique using the ProteinChip array-based SELDI-TOF-MS analysis for high throughput profiling of phosphoproteins/phosphopeptides in human serum for the early detection and diagnosis as well as for the molecular staging of human cancer. The methodology and experimental approach consists of five steps: (1) generation of a total peptide pool of serum proteins by a global trypsin digestion; (2) rapid isolation of phosphopeptides from the total serum peptide pool by an affinity selection, purification, and enrichment using a novel automated micro-bioprocessing system with phospho-antibody-conjugated paramagnetic beads and a hybrid magnet plate; (3) high throughput phosphopeptide analysis on ProteinChip arrays by automated SELDI-TOF-MS; and (4) bioinformatics and statistical methods for data analysis. This method with appropriate modifications may be equally applicable to serine-, threonine- and tyrosine-phosphorylated proteins and for selectively isolating, profiling, and identifying phosphopeptides present in a highly complex phosphor-peptide mixture prepared from various human specimens such as cells, tissue samples, and serum and other body fluids.

Key words: Phosphoprotein, Phosphopeptide, Phosphoproteome, High throughput Phospoprotein/Peptide Profiling, ProteinChip Arrays, SELDI-TOF-MS

1. Introduction

The reversible protein phosphorylation is a key regulating switch that controls a wide range of biological functions and activities (1–8). Particularly, phosphorylation of protein kinases plays a critical role in signaling pathways involved in oncogenesis and pathogenesis of various human cancers (6, 8–10). It is clear that the abnormal protein phosphorylation is associated with many major diseases, such as diabetes, rheumatoid arthritis, cardiovascular disease, and cancers (1, 10, 11). Therefore it is vitally important to understand the intracellular signaling events that control protein phosphorylation. Many toxins that are known to cause cancers act by affecting the functions of kinases or phosphatases (5, 9–12). Although protein kinases are now one of the major groups of proteins being targeted by drug discovery and therapeutics programs, the substrates, or proteins that these kinases and phosphatases affect, are largely unknown. Thus, determination of the sites and magnitudes of protein phosphorylation has been an essential step in the analysis of the control of many biological systems. A high throughput analysis of phosphorylation of serum proteins would provide a simple, logical, and useful tool for a functional diagnosis and prediction of human cancers in association with these important molecular events (5, 8, 9, 13–16). Moreover, the current advanced cancer treatment with anti-angiogenesis agents and protein kinase inhibitors showed profound impact on phosphorylation/dephosphorylation of proteins involved in cell proliferation, apoptosis, angiogenesis, tumor progression, and metastasis in laboratory and in preclinical and clinical settings (10, 11, 17). These phospho-modified proteins are secreted from cells and into the circulation system and are easily available in the serum. Thus, a high-throughput proteomic profiling of phosphopeptides in serum samples exposed to these agents will allow the identification of specific serological biomarkers associated with their biological action (3, 8, 14, 17–20). However, direct determination of phosphorylation of individual proteins in a biological system has been difficult to date (5, 9, 14, 15, 21). It typically requires the purification to homogeneity of the phosphoprotein of interest before analysis and there is currently no method available to study this aspect in detail and consistently. Thus, there has been a substantial need for a more rapid, global, and general method for the analysis of protein phosphorylation in complex protein mixtures (3, 14, 17–20, 22). In addition, a large-scale global phospho-proteome analysis poses challenges in several fronts including to simultaneously isolate and enrich phosphopeptides in several hundred parallel samples without introduction of significant experimental errors and to maintain consistent integrity of proteome among all the samples (2, 3, 14, 15, 18, 19, 21, 23). To address these challenges, we developed an innovative, rapid, and simplified method for serum phoshopeptide separation and enrichment using a one-step affinity capture of phosphopeptides on phospho-antibody-conjugated paramagnetic beads or nanoparticles and separation by a hybrid magnet. This method offers the advantage of automation to avoid human errors and enable a high throughput serum phosphopeptide preparation, which is readily coupled with an automated peptide profiling and analysis on ProteinChip arrays by SELDI-TOF-MS.

2. Materials

2.1. Reagents

1. Phospho-tyrosine antibody: the phospho-tyrosine mouse mAbs (P-Tyr-100) developed by Cell Signaling Technology (Danvers, MA) is a high affinity mouse monoclonal antibody and provides an exceptionally sensitive new tool with increased utility for studying tyrosine phosphorylation and monitoring tyrosine kinase activity in high throughput tyrosine phosphor-protein/peptide analysis. The antibody is supplied in 10 mM sodium HEPES (pH 7.5), 150 mM NaCl, 100 μg/ml BSA and 50% glycerol. Store at −20°C.

2. Surface Activated Dynabeads: the Dynabeads® MyOne™ Tosylactivated (Invitrogen, Carlsbad, CA) are superparamagnetic and uniformpolystyrene beads (1.0 μm in diameter) coated with a polyurethane layer. The dynabeads are used for conjugation with phosphor-antibodies and for biomagnetic separation and enrichment of phosphor-antibodycaptured phosphor-peptides.

3. Sequencing Grade Modified Trypsin: Sequencing Grade Modified Trypsin (Promega, Madison, WI) is a porcine trypsin modified by reductive methylation, rendering it resistant to proteolytic digestion (24). Sequencing Grade Modified Trypsin is supplied as lyophilized powder and can be reconstituted in 50 mM acetic acid. The substrate is dissolved in 50 mM Tris–HCl (pH 7.6), 1 mM $CaCl_2$, and the enzyme is diluted in 50 mM acetic acid.

4. Human Serum: human clinical serum samples are collected and prepared using approved clinical protocols and standard methods. Serum samples are stored at −80°C.

2.2. Magnetic Plate

A new class of hybrid magnet plates has recently been developed at the Joint Genome Institute and Lawrence Berkeley National Laboratroy (JGI/LBNL, Berkeley, CA) for high throughput purification of biological samples for functional genomics and proteomics and for affinity drug screening due to its superior capability of selectively separating proteins and DNA from complex biological mixtures based on a magnetic field (25, 26). These magnet plates are ideal for any process that requires automated bead manipulation in high-density microtiter plates containing sample volumes in a range of 3–300 μl. The novel hybrid magnetic structure combines a permanent magnet with ferromagnetic materials that produces magnetic fields significantly higher than those of any commercially available magnetic plate. More importantly, the fields at a distance of 1 cm above the magnet are more than 1,000-fold stronger than those of the commercial 96-well magnet. This feature allows for more vigorous washing and sample recovering. The second generation 96-well hybrid magnet plate has been designed and constructed for our proteomics platform by physicists and engineers at JGI/LBNL, producing fields well above 10,000.0 G, which allows a more efficient separation of affinity-captured phosphopeptides from crude serum peptide mixtures and thus improves reproducibility and sensitivity of proteomic analysis by reducing processing loss and increasing peptide recovery while retaining a high peptide-proteome integrity. Alternatively, the commercially available 96-well magnetic plates can also be used.

2.3. ProteinChip Arrays and Peptide Standards

1. SEND-ID ProteinChip arrays (Bio-Rad, Hercules, CA) have C18 as a functional group and are used for phosphopeptide profiling and fingerprinting on SELDI-TOF-MS. IMAC30 ProteinChip arrays (Bio-Rad) can also be used for phosphopeptide analysis.
2. All-in-1 Peptide standards (Bio-Rad) for SELDI-TOF-MS is supplied as a dry powder in a glass vial with rubber stopper and is freshly reconstituted in Reconstitution Solution.

2.4. Buffers and Solutions

1. Trypsin Resuspension Buffer: 50 mM acetic acid.
2. Protein Denaturation Solution: 6 M guanidine HCl (or 6–8 M urea), 50 mM Tris–HCl (pH 8), 2–4 mM DTT (or β-mercaptoethanol).
3. 1× PBS (phosphate buffered saline) (pH 7.4): 0.26 g $NaH_2PO_4 \times H_2O$ (MW 137.99), 1.44 g $Na_2HPO_4 \times 2H_2O$ (MW 177. 99), 8.78 g NaCl (MW 58.5). Dissolve in 900 ml dH_2O. Adjust pH to 7.4 and the volume to 1,000 ml.
4. 10× Phosphate Buffered Saline (PBS): To prepare 1 L add 80 g sodium chloride (NaCl), 2 g potassium chloride (KCl), 14.4 g sodium phosphate, dibasic (Na_2HPO_4), and 2.4 g potassium phosphate, monobasic (KH_2PO_4) to 1 L dH_2O. Adjust pH to 7.4.
5. Coating buffer: 0.1 M sodium borate buffer (pH 9.5): 6.183 g H_3BO_3 (MW 61.83). Dissolve in 800 ml distilled water. Adjust pH to 9.5 using 5 M NaOH and adjust volume to 1,000 ml with distilled water. The coating buffer is used for pre-washing and coating of Dynabeads.
6. 3 M ammonium sulphate stock solution: 39.6 g $(NH_4)_2SO_4$ (MW 132.1). Dissolve in 0.1 M sodium borate buffer (pH 9.5), adjust pH to 9.5 and adjust volume to 100 ml.
7. Blocking buffer: PBS pH 7.4 with 0.5% (w/v) BSA and 0.05% Tween 20 in 100 ml PBS. Blocking buffer is used for blocking of all precoated Dynabeads. Do not use this buffer or any buffer containing protein or amino-groups (glycine, Tris etc.) for pre-washing or coating of Dynabeads.
8. Washing buffer: PBS pH 7.4 with 0.1% (w/v) BSA and 0.05% Tween-20 in 100 ml PBS. If a preservative is needed in the

coated product, a final concentration of 0.02% (w/v) sodium azide (NaN$_3$) may be added to washing buffer. This preservative is cytotoxic and must be carefully removed before use by washing.

9. Elution buffers: Any conventional method for protein elution can be used, e.g., 0.1 M citrate pH 3, 0.1 M glycine–HCl pH 2.5, or 0.1 M glycine–NaOH pH 10. All reagents used should be analytical grade. Organic solvent containing 50% acetonitrile (CAN) and 0.1% Triflouroaacetic acid (TFA) (Sigma, St. Louis, MO) is best for the SEND-ID chip.

10. Peptide Standards Reconstitution Solution: 10 mM ammonium acetate, 25% acetonitrile, and 1.25% trifluoroacetic acid.

3. Methods

An important goal of clinical proteomics is to develop sensitive, specific, and robust proteomic platforms to simultaneously measure the human proteome in clinically relevant specimens and to establish protein signature profiles for discriminating between the normal and disease states (27–35). Serum potentially carries a rich archive of histological and biological information and is attracting increasing interest in clinical proteomics. A throughput profiling and an accurate measurement of these serum proteome would serve to improve early detection, diagnosis, and prognosis of cancers and identify new therapeutic targets (27–35). While the importance of studying the serum proteome is obvious, the characterization and analysis of serum proteins, however, are analytically challenging due to their extremely high dynamic range of concentration that spans more than 9–10 orders of magnitude and to the complexity that is composed of biomolecules ranging from large proteins and lipids to small metabolite hormones, peptides, amino acids, and electrolytes. Particularly, the serum protein contents are dominated by a handful of abundant proteins such as albumin, immunoglobulins, haptoglobin, transferrin, and lipoproteins that account for more than 99% of serum protein masses and overwhelmingly shadow the detection of those low abundant but biologically important molecules (30). The reduction of sample complexity and depletion of the level of these abundant proteins are essential first steps for a successful and efficient analysis of the serum proteome. Affinity depletion methods have therefore been developed to remove abundant serum proteins such as albumin and immunoglobulin from serum prior to mass spectrometric analysis (36–38). One of the major pitfalls of these protein depletion methods, however, is that many important low molecular weight proteins or peptides can be concomitantly removed during the sample processing as well (27, 28, 30). Classical methods such as sample fractionation and purification by liquid chromatography with various media, separation by gel electrophoresis, sample desalting and concentration by dialysis, centrifugation, and immuno-precipitation, are often labor-intensive and demand large quantities of sample, suffer from attendant analyte loss due to non-specific binding and dilution effects, and easily introduce experimental errors during multi-task and multi-step sample preparation thus hampering sample quantification and parallel comparison (30, 38, 39). A simple, direct, and efficient mass spectrometric sample preparation and protein/peptide detection in heterogeneous samples is much needed.

In this method, we use innovative antibody-conjugated magnetic beads to specifically capture a subset of biologically important phosphopeptides from the trypsin-digested serum peptide mixtures. The captured peptides are rapidly and efficiently separated and purified by a novel hybrid magnet specifically developed for our proteomic analysis. This method will significantly reduce the complexity of serum proteins, completely eliminate the interference of abundant serum proteins but not affect the integrity of serum proteome, and avoid multi-steps of serum sample fractionation and purification, thus, allowing a high throughput sample preparation, quantification, and parallel comparison. A comparison of technologies and applications between the existing proteomic approaches (Method 1) and our new serum phosphopeptide proteomic platform (Method 2) is shown in Fig. 1.

The ProteinChip-based SELDI technology is currently being used to successfully detect disease-associated proteins in complex biological specimens and has primarily been applied to search for the cancer-relevant biomarkers in clinical serum samples (33, 34, 40–45). These studies emphasize the capability and potential of SELDI for the detection and characterization of differentially expressed proteins or proteomic patterns for detection and prediction of diseases. For serum proteomics to realize its full potential, however, several potential problems and controversy in sensitivity and reproducibility concerning the SELDI profiling approach needs to be addressed (32, 34, 46, 47). Semmes et al. (48) has recently assessed the platform reproducibility using SELDI-mediated serum protein profiling for the detection of prostate cancer and demonstrated the reproducibility of SELDI serum profiling between laboratories and suggested that this approach could provide a reproducible diagnostic assay platform. The most severe limitation for the reproducibility may be a result of loss of the majority of proteins and peptides present in the sample while using SELDI for the protein profiling (27, 28, 43, 48). This in turns leads to the rather low-resolution pattern that represents only a minority of proteins and peptides in the serum. Other limitations for SELDI protein profiling are due to variability such as in

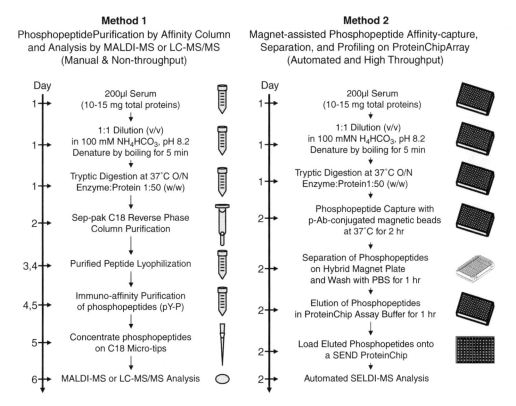

Fig. 1. Comparison of methods and platforms for phosphoprotein/phosphopeptide proteomics analysis by mass spectrometry. Method 1, the conventional phosphopeptide purification by affinity column and analysis by MALDI-MS or LC-MS/MS. Method 2, the magnet-assisted phosphopeptide affinity-capture, separation, enrichment, and profiling on ProteinChip arrays by SELDI-TOF-MS.

instrumental laser desorption energy level and ProteinChip array quality and in sample collection, processing, and storage (32, 49–53). These changes resulted in reproducible changes in serum proteome, and can sometimes overshadow the biological changes in the serum samples (32). In this study, we use SELDI-TOF-MS to profile serum phosphopeptide sub-proteome instead of total proteins and, thus, will partially circumvent these problems and enhance reproducibility because it concentrates all measurements in a small region of low mass peptides within 200–5,000 Da with an extremely high mass resolution by SELDI-TOF-MS spectrometry. Furthermore, because the affinity capture is performed after a complete global proteolytic digestion of serum proteins, phosphopeptides cleaved from larger proteins are equally captured together with other phosphopeptides in the pool and can be easily and precisely detected on MS spectroscopy, therefore circumventing the difficulty and inefficiency of using mass spectrometry such as SELDI or MALDI in detection of large protein species (>30–50 kDa), improve the measurement sensitivity, and gain a higher coverage of serum proteome.

3.1. Preparation of Phospho-Antibody-Conjugated Magnetic Beads

The Tosylactivated Dynabeads® (1–2 μm in diameter) (Invitrogen, Carlsbad, CA) is used as a solid phase to conjugate phosphor-antibodies for a biomagnetic separation of phosphopeptides from the trypsin-digested serum peptide mixtures. The general protocols given below are based on experience with either phospho-tyrosine (pY)-specific antibodies (Cell Signaling Technology) or phospho-Serine/Threonine (pS/pT)-specific antibodies (BD Biosciences). When incubating the beads with the ligand of choice, it will be physically adsorbed onto the surface of Dynabeads MyOne™ Tosylactivated first and followed by the formation of covalent bonds over time (see Note 1).

1. The volume of beads used is based on the number of serum samples to be analyzed (1 mg beads per 200 μl serum sample). The conjugation is at a ratio of 40 μg antibodies to 1 mg beads (w/w). These conditions are for the coating of 1.0 mg phosphor-antibody to 25 mg Dynabeads MyOne™ Tosylactivated (250 μl at 100 mg beads/ml).
2. Resuspend Dynabeads thoroughly by vortexing for 30 min. Transfer 250 μl of beads into a test tube. Place the tube on a magnet (Dynal MPC, Invitrogen) for 2 min or until the beads have migrated to the side of the tube and the liquid is clear.
3. Pipette the supernatant off carefully, leaving beads undisturbed.
4. Remove the test tube from the magnet and resuspend the beads thoroughly in 1 ml of Coating buffer (Solution 5) by vortexing.
5. Repeat steps 3–4.
6. Resuspend the washed beads in 100 μl volume coating buffer and the beads are ready for conjugation with antibodies.
7. Dilute 1.0 mg of antibodies in coating buffer to a total of 600 μl. For optimized coating, the antibodies may be pre-treated and acidified (see Note 2)
8. Add 970 μl of coating buffer to the above washed beads (100) and mix properly.
9. Add the diluted antibodies (1.0 mg/600 μl) to the suspended beads and mix properly.
10. Add 830 μl of 3 M ammonium sulphate stock solution to the antibody/beads mixture to a total of 2,500 μl of conjugation reaction.
11. Incubate the conjugation reaction for 16–24 h at 37°C with slow tilt rotation. Do not let the beads settle during the incubation period (see Note 3 for optimized coating time, temperature, and pH).

12. After incubation, place the tube on the magnet for 2 min, or until the beads have migrated to the side of the tube, and remove the supernatant.
13. Add the same total volume (2,500 μl) of PBS with 0.5% BSA and 0.05% Tween-20 and incubate at 37°C over night.
14. Wash three times with PBS with 0.1% BSA and 0.05% Tween-20, and resuspend the washed conjugates to the desired volume or concentration. The Dynabeads are now coated and ready for use (see Note 4).
15. For storage, the desired preservative, e.g., 0.02% sodium azide may be added and store at 2–8°C. The coated beads can usually be stored for several months at 2–8°C, depending on the stability of your immobilized antibodies.

3.2. Protease Digestion of Serum Proteins

1. Serum Protein Denaturation: Serum proteins require denaturation and disulfide bond cleavage before enzymatic digestion can go to completion. 200 μl (10–15 mg of total proteins) of serum is used for each individual assay. Dilute serum sample 1:1 (v/v) in 100 mM NH_4HCO_3, pH 8.2, in each well of a 96-well plate.
2. Boil for 5 min in a heating-block that fits a 96-well plate. If smaller amounts of protein are to be digested, the recommended conditions given can be scaled down proportionally. However, under no conditions should less than 25 μl of dissolving agent be used.
3. After denaturation, allow the reaction to cool to room temperature.
4. Protease Digestion: add sequencing grade modified trypsin (Promega) to a final protease:serum proteins ratio of 1:50 (w/w). Incubate at 37°C overnight.
5. Remove a small aliquot and chill the reaction on ice or freeze. Add an inhibitor to the aliquot to terminate the protease activity or precipitate the sample by the addition of TCA to a 10% final concentration.
6. Determine the extent of digestion by subjecting a portion of the digestion products to reverse phase HPLC or SDS-PAGE. If further proteolysis is required, return the reaction tube to 37°C and continue incubating until the desired digestion is obtained.
7. The reaction can be terminated by freezing or by the addition of specific inhibitors (see Note 5).

3.3. Capture, Separation, and Enrichment of Serum Phosphopeptides

1. For affinity capture and enrichment of serum phosphopeptides, wash the phosphor-antibody-conjugated magnetic beads with two volumes of 1× PBS three times to remove any unbound antibodies.

2. Resuspend the phosphor-antibody (p-Ab)-beads conjugates in PBS at a desired concentration (1 mg per 200 μl of serum sample).

3. Transfer the p-Ab/beads conjugates at a ratio of 200 μl trypsin-digested serum proteins to 1 mg p-Ab/beads into each well of the 96-well plates containing digested serum proteins.

4. Incubate the plate for 1 h at room temperature with gentle shaking.

5. Place the plate on a 96-well magnet plate and allow phosphor-peptide-captured beads to settle down.

6. Gently remove the supernatant and the unbound peptides by washing three times with 400 μl of PBS.

7. The captured phosphor-peptides are then eluted in 15 μl of elution buffer comprising 50% acetonitrile (CAN) and 0.1% Trifluoroacetic acid (TFA).

8. Place the plate on a 96-well hybrid magnet plate. After the beads settle down, carefully collect the eluted phosphor-peptides into a fresh 96-well plate.

9. All these steps of process are performed automatically using a Biomek-2000 Laboratory Automation Workstation (Beckman, Fullerton, CA).

3.4. Phosphopeptide Profiling on ProteinChip Arrays by SELDI-TOF-MS

The enriched phosphopeptides are further processed automatically using a micro bio-processor (Bio-Rad) and a Biomek-2000 Laboratory Automation Workstation (Beckman, Fullerton, CA) and loaded onto a SEND (Surface Enhanced Neat Desorption) or IMAC proteinchip array purchased from (Bio-Rad). SEND arrays are unique compared to other arrays by having the energy-absorbing molecule (EAM) incorporated into the array chemistry and is added after sample addition. They are best suited for small molecule analysis, which in our case are phosphopeptides. Pooled serum samples are used as quality assurance (QA) controls and samples are randomly loaded onto each ProteinChip array with duplicates. One spot is loaded on each array with peptide standards for peptide mass calibration. The chips are automatically loaded and analyzed by SELDI-TOF-MS spectrometer (Model PSII or new model PCS 4000, Bio-Rad).

1. Add 5 μl of 0.1% TFA to each spot of a SEND-ID array and then quickly remove.

2. Add 10 μl of the eluted phosphor-peptides onto each spot of the SEND-ID array that is placed in a humidity chamber and incubate for 30 min.

3. Remove samples from the array surface and wash bound phospho-peptides quickly once with 5 μl of 0.1% TFA.

4. Then add 2 μl of 25% ACN and 0.1% TFA to each spot. Air-dry.

5. Read arrays with desired and optimized instrumental parameters and settings in a ProteinChip reader, according to manufacturer's instruction.

3.5. Data Processing and Analysis

The SELDI-MS spectral data collected from the mass spectrometer are calibrated and subjected to further analysis using bioinformatics tools and statistical methods developed by Dr. Coombes (52, 54). Advanced proteomic data processing and analysis methodologies and bioinformatics algorithms are needed to address concerns regarding reliability, sensitivity, and reproducibility of peak detection, quantification, and identification in clinical serum proteomics (32). Some advanced methodologies and algorithms for spectral alignment, baseline correction and normalization, peak detection and quantification, and statistical analysis of peaks are also needed for evaluating clinical significance of cancer diagnostic peptides (32, 49–55). These methods will improve the reproducibility of peak quantifications and provide tools for evaluating the variations in this phosphor-peptide proteomics platform and more accurately interpret serum phosphor-peptide profiling results.

To overcome the technology barriers and circumvent the potential problems and limitations facing MS spectrometry-based serum proteomics and as demonstrated in the above serum protein-profiling experiments, we have developed an innovative and integrated functional proteomics technique using the ProteinChip array-based SELDI-MS analysis for a high throughput profiling of phosphopeptides in human serum. To demonstrate the feasibility of this proteomic platform for clinical serum sample processing and analysis, we performed phosphopeptide (phosphor-Tyrosine) proteomic profiling on serum samples from human normal and lung cancer patients in different stages and with different smoking histories. We analyzed three groups of samples: (1) 20 serum samples from lung cancer patients collected by Dr. Roth in the Department of Thoracic & Cardiovascular Surgery (Cancer Group 1, C-G1), (2) 20 lung cancer serum samples collected by Dr. Spitz in the Department of Epidemiology (Cancer Group 2, C-G2), and (3) 20 serum samples from normal controls. Tyrosin-phosphopeptides (pYPs) were prepared and selectively isolated from these serum samples as described in the Subheading 3. The purified pYPs were randomly loaded onto SEND ProteinChip arrays in duplicate. Phosphopeptide MS-spectra and data were analyzed using bioinformatics tools and statistical methods developed by Dr. Coombes (52, 54).

We used wavelets and the mean spectrum for peak detection. Briefly, we first computed the mean of the aligned, baseline-corrected, normalized spectra. We used an undecimated discrete wavelet transform (UDWT) to denoise the mean spectrum by hard thresholding the wavelet coefficients that were less than ten times the standard deviation. Peaks were defined as local maxima in the denoised mean spectrum. Along with the location of each peak, we also recorded an interval that contained the peak by finding the nearest local minimum on either side of the peak. Using this procedure, we detected 622 distinct peaks spanning a m/z range from 50 to 5,500 Da (Fig. 2a). The smallest signal-to-noise ratio (S/N) of any of these peaks was 4.92; the median S/N was 416. In order to quantify the peaks in the individual spectra, we began by locating

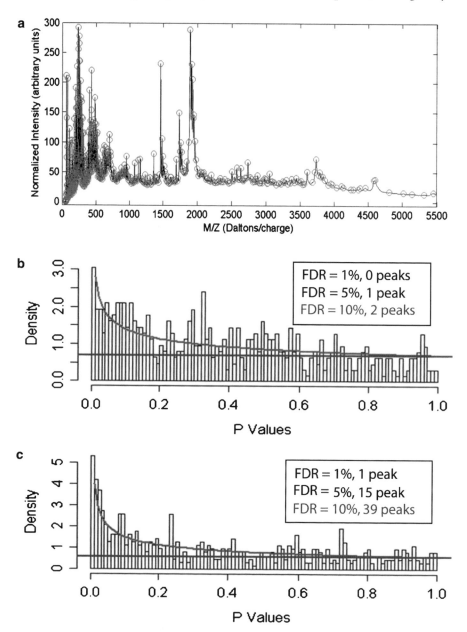

Fig. 2. Detection, quantification, and statistical analysis of specific human serum phosphopeptide peaks determined by ProteinChip array-based SELDI-TOF-MS. (a) Phosphopeptide peaks are detected using wavelets and determined by the mean spectrum of the aligned, base-line-corrected, and normalized spectra. More than 600 phospho-tyrosine peptide peaks are detected from the magnet-assisted and affinity-enriched human serum phosphopetide pools. Statistical analysis of

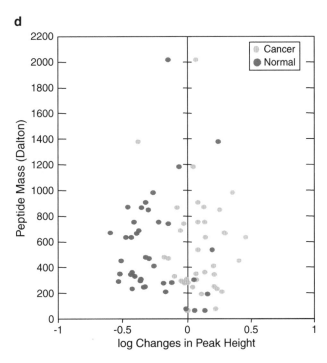

phosphor-peptide modulations between the normal and lung cancer serum samples using a beta-uniform mixture (BUM) Fig. 2. (continued) model to estimate the false discovery rate (FDR). (**b**) BUM analysis of peaks that are different between the three groups of samples using one-way NOVA, and (**c**) BUM analysis of *p*-values from an *F*-test of the significance of group effects after accounting for other technological factors, including laser intensity, pressure, and spot position. Peaks with significant changes ($p<0.05$) between the normal and the cancer groups are determined at FDR = 1, 5, and 10%, respectively. (**d**) Scattered plot of 39 pairs of phosphor-tyrosine peptides with significant changes between the normal and lung cancer serum samples, as determined by BUM in **c**.

the time interval containing the peak. We then took the maximum value of the spectrum in that interval and subtracted the three minimum values in the interval to define the peak height. Peak quantification was performed on the aligned, baseline-corrected, normalized spectra. Implicitly, the minimum value was used as a local estimate of the baseline in the interval. Because no smoothing was performed, the peak heights might be slightly biased on the high side. However, this is a reasonable trade-off because it decreases the variance.

Information on the peak locations and heights was further exported from MATLAB and imported to the software program R for statistical analysis. To answer the primary question whether there are any peaks that are different between cancer samples and normal samples, we performed a one-way analysis of variance (ANOVA) using a single factor that takes on three levels (Cancer-G1 Cancer-G2, verses Normal). We performed a separate ANOVA for each of the 622 peaks, using the base-two logarithm of the peak height to try to separate the three sample groups. For each peak, we recorded the *p*-value from an *F*-test of the model; small *p*-values suggest that the peak height is different between at least two of the three groups in the study. In order to account for multiple testing, we modeled the set of *p*-values using a beta-uniform mixture (BUM) model to estimate the false discovery rate (FDR) (56–59). Setting FDR at 1, 5, and 10%, we found 0, 1, and 2 significant peaks, respectively (Fig. 2b). However, using a BUM analysis after accounting with other technological factors including laser intensity, vacuum chamber pressure, and spot positions on the ProteinChip arrays, we found 1, 15, and 39 significant peaks with FDR= 1, 5, and 10%, respectively (Fig. 2c). The fold changes in intensity detected on MS profiles between the normal and cancer serum samples are plotted in Fig. 2d. Using this phosphopeptide proteomic profiling technology and data analysis we are able to find peaks that significantly differ between the three groups. The data generated from these phosphopeptide profiles are also highly reproducible, as shown by the consistent mass spectra among each sample group (Fig. 3). Differences in sample handling explain

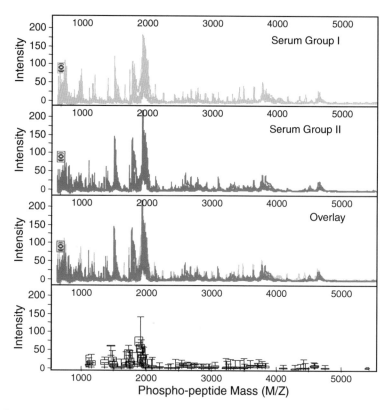

Fig. 3. Profiles and analysis of tyrosine-specific phosphopeptides in normal and lung cancer serum samples on SEND ProteinChip arrays by SELDI-TOF-MS. The variations of phosphopetide levels as defined by peak intensity on the mass spectra among serum samples are shown by the overlapping spectra (*top three panels*) and by the *box plots* (*bottom panel*) with *error bars* indicating the ranges of each paired peptide peak among three serum sample groups.

some of the peaks that are found to be differentially expressed. Nevertheless, many of the changes can clearly be attributed to differences between normal samples and cancer samples, regardless of who collected them. These pilot experiments clearly demonstrate the feasibility of our serum phosphopetide proteomic platform in detecting temporal changes of phosphopeptide proteome in clinically relevant serum samples. Our proteomics platform also demonstrated the capability of overcoming and circumventing a number of technological problems and barriers facing the current MS spectrometry-based proteomics technologies, including reduction of proteome complexity, enhancement of specificity and sensitivity of detection of low abundance proteins and peptides, increase of throughput rate of sample process and analysis, and improvement of identification and quantification of specific peptides and proteins on MS profiles and proteomic data processing and analysis.

4. Notes

1. The efficacy of immunomagnetic separation is critically dependant on the specificity and avidity of the antibody or other ligand applied. A concentration of 40 µg antibody/mg Dynabeads is generally optimal for coating. Antibody/protein to be coated directly onto the surface of Dynabeads must be purified, since all proteins will bind to the bead surface. Sugars or stabilizers may disturb the binding and should be removed from the antibody preparation.

2. For antibody pre-treatment and acidification, in general, lowering pH to 2.5 for 15 min at room temperature or 1 h at 1–4°C, and then raising the pH to approximately neutral prior to addition of the beads, will increase binding and function of antibodies, but this must be optimized for your specific antibodies.

3. The physical adsorption to the bead surface is rapid, while the formation of covalent bonds will need more time. After the recommended 16–24 h at 37°C, a maximal chemical binding is achieved. Coating at 20°C will require an extended incubation time to 48 h and longer to obtain the same degree of chemical binding. At 4°C the chemical binding is very slow (>48 h). Both higher temperatures and a higher pH will speed up the formation of covalent bonds, provided that the antibodies in question are stable and functional under these conditions. Sodium borate buffer pH 9.5 is recommended. Molarities between 0.1 and 0.5 are optimal.

4. If the presence of BSA will interfere with your downstream application, this protein can be omitted from the buffer. Detergent may similarly be omitted.

5. Trypsin can also be inactivated by lowering the pH of the reaction to below 4. Trypsin will regain activity as the pH is raised above 4. Reducing the temperature will decrease the digestion rate. Longer incubation periods, up to 24 h, may be required depending on the nature of the protein.

Acknowledgments

The authors would like to thank Drs. Xifeng Wu and Margaret Spitz for providing lung cancer serum samples for serum phosphopeptide profiling and analysis and Dr. Kevin Coombes for bioinformatics and statistical analysis, at The University of Texas M. D. Anderson Cancer Center, Houston, TX, and David Humphries at Lawrence Berkeley National Laboratory, Oak Land, CA for developing hybrid magnetic plates for serum phosphopeptide enrichment. This work was partially supported by grants from NIH/NCI SPORE P50CA070907, RO1CA116322, and MMHCC U01CA105352; DOD PROSPECT W81XWH-0710306; The University of Texas M. D. Anderson Cancer Center Support Core Grant (CA16672).

References

1. Hunter, T. (2000) Signaling–2000 and beyond. *Cell.* **100**, 113–127.
2. Oda, Y., Huang, K., Cross, F.R., Cowburn, D., and Chait, B.T. (1999) Accurate quantitation of protein expression and site-specific phosphorylation. *Proc. Natl. Acad. Sci. USA.* **96**, 6591–6596.
3. Oda, Y., Nagasu, T., and Chait, B.T. (2001) Enrichment analysis of phosphorylated proteins as a tool for probing the phosphoproteome. *Nat. Biotechnol.* **19**, 379–382.
4. Peters, E.C., Brock, A., and Ficarro, S.B. (2004) Exploring the phosphoproteome with mass spectrometry. *Mini-Rev. Med. Chem.* **4**, 313–324.
5. Johnson, S.A. and Hunter, T. (2005) Kinomics: methods for deciphering the kinome. *Nat. Methods.* **2**, 17–25.
6. York, J.D. and Hunter, T. (2004) Signal transduction. Unexpected mediators of protein phosphorylation. *Science.* **306**, 2053–2055.
7. Johnson, S.A. and Hunter, T. (2004) Phosphoproteomics finds its timing. *Nat. Biotechnol.* **22**, 1093–1094.
8. Manning, G., Whyte, D.B., Martinez, R., Hunter, T., and Sudarsanam, S. (2002) The protein kinase complement of the human genome. *Science.* **298**, 1912–1934.
9. Rush, J., Moritz, A., Lee, K.A., Guo, A., Goss, V.L., Spek, E.J. et al. (2005) Immunoaffinity profiling of tyrosine phosphorylation in cancer cells. *Nat. Biotechnol.* **23**, 94–101.
10. Hunter, T. (2002) Tyrosine phosphorylation in cell signaling and disease. *Keio J. Med.* **51**, 61–71.
11. Hunter, T. (1998) The role of tyrosine phosphorylation in cell growth and disease. *Harvey Lectures.* **94**, 81–119.
12. Alonso, A., Sasin, J., Bottini, N., Friedberg, I., Friedberg, I., Osterman, A. et al. (2004) Protein tyrosine phosphatases in the human genome. *Cell.* **117**, 699–711.
13. Hornbeck, P.V., Chabra, I., Kornhauser, J.M., Skrzypek, E., and Zhang, B. (2004) PhosphoSite: A bioinformatics resource dedicated to physiological protein phosphorylation. *Proteomics.* **4**, 1551–1561.
14. Gerber, S.A., Rush, J., Stemman, O., Kirschner, M.W., and Gygi, S.P. (2003) Absolute quantification of proteins and phosphoproteins from cell lysates by tandem MS. *Proc. Natl. Acad. Sci. USA.* **100**, 6940–6945.
15. Steen, H., Jebanathirajah, J.A., Rush, J., Morrice, N., and Kirschner, M.W. (2006)

Phosphorylation analysis by mass spectrometry: myths, facts, and the consequences for qualitative and quantitative measurements. *Mol. Cell. Proteomics.* **5**, 172–181.

16. Pan, S., Zhang, H., Rush, J., Eng, J., Zhang, N., Patterson, D. et al. (2005) High throughput proteome screening for biomarker detection. *Mol. Cell. Proteomics.* **4**, 182–190.

17. Brill, L.M., Salomon, A.R., Ficarro, S.B., Mukherji, M., Stettler-Gill, M., and Peters, E.C. (2004) Robust phosphoproteomic profiling of tyrosine phosphorylation sites from human T cells using immobilized metal affinity chromatography and tandem mass spectrometry. *Anal. Chem.* **76**, 2763–2772.

18. Mann, M., Ong, S.E., Gronborg, M., Steen, H., Jensen, O.N., and Pandey, A. (2002) Analysis of protein phosphorylation using mass spectrometry: deciphering the phosphoproteome. *Trends Biotechnol.* **20**, 261–268.

19. Salomon, A.R., Ficarro, S.B., Brill, L.M., Brinker, A., Phung, Q.T., Ericson, C. et al. (2003) Profiling of tyrosine phosphorylation pathways in human cells using mass spectrometry. *Proc. Natl. Acad. Sci. USA.* **100**, 443–448.

20. Zhou, H., Watts, J.D., and Aebersold, R. (2001) A systematic approach to the analysis of protein phosphorylation. *Nat. Biotechnol.* **19**, 375–378.

21. Zhang, H., Zha, X., Tan, Y., Hornbeck, P.V., Mastrangelo, A.J., Alessi, D.R. et al. (2002) Phosphoprotein analysis using antibodies broadly reactive against phosphorylated motifs. *J. Biol. Chem.* **277**, 39379–39387.

22. Rush, J., Moritz, A., Lee, K.A., Guo, A., Goss, V.L., Spek, E.J. et al. (2005) Immunoaffinity profiling of tyrosine phosphorylation in cancer cells. *Nat. Biotechnol.* **23**, 94–101.

23. Mann, M. (1999) Quantitative proteomics? *Nat. Biotechnol.* **17**, 954–955.

24. Rice, R.H., Means, G.E., and Brown, W.D. (1977) Stabilization of Bovine Trypsin by Reductive Methylation. *Biochim. Biophys. Acta.* **492**, 316–321.

25. Elkin, H., Kapur, T., Humphries, D., Pollard, M., Nammon, N., and Hawkins, T. (2002) Magnetic bead purification of labeled DNA fragments for high throughput capillary electorphoresis sequencing. *Biotechniques.* **32**, 1296–1307.

26. Humphries, D.E., Pollard, M.J., and Elkin, C.J. High Performance Hybrid Magnetic Structure for Biotechnology Applications. The Regents of the University of California. **10/305658** (6954128B2). 2002 California, USA (Patent).

27. Aebersold, R., Rist, B., and Gygi, S.P. (2000) Quantitative proteome analysis: methods and applications. *Annal. New York Acad. Sci.* **919**, 33–47.

28. Aebersold, R. and Mann, M. (2003) Mass spectrometry-based proteomics. *Nature.* **422**, 198–207.

29. Anderson, N.L., Matheson, A.D., and Steiner, S. (2000) Proteomics: applications in basic and applied biology. *Curr. Opin. Biotechnol.* **11**, 408–412.

30. Anderson, N.L. and Anderson, N.G. (2002) The human plasma proteome: history, character, and diagnostic prospects. *Mol. Cell. Proteomics.* **1**, 845–867.

31. Celis, J.E.G. (2003) Proteomics in translational cancer research: Toward an integrated approach. *Cancer Cell.* **3**, 9–15.

32. Coombes, K.R., Morris, J.S., Hu, J., Edmonson, S.R., and Baggerly, K.A. (2005) Serum proteomics profiling–a young technology begins to mature. *Nat. Biotechnol.* **23**, 291–292.

33. Petricoin, E.F., Zoon, K.C., Kohn, E.C., Barrett, J.C., and Liotta, L.A. (2002) Clinical proteomics: translating benchside promise into bedside reality. *Nat. Rev. Drug Discov.* **1**, 683–695.

34. Petricoin, E.F. and Liotta, L.A. (2004) SELDI-TOF-based serum proteomic pattern diagnostics for early detection of cancer. *Curr. Opin. Biotechnol.* **15**, 24–30.

35. Sauer, S., Lange, B.M., Gobom, J., Nyarsik, L., Seitz, H., and Lehrach, H. (2005) Miniaturization in functional genomics and proteomics. *Nat. Rev. Genet.* **6**, 465–476.

36. Shrivastava, A., von Wronski, M.A., Sato, A.K., et al. (2005) A distinct strategy to generate high-affinity peptide binders to receptor tyrosine kinases. *Prot. Engin., Design. Select.* **18**, 417–424.

37. Sato, A.K., Sexton, D.J., Morganelli, L.A., et al. (2002) Development of mammalian serum albumin affinity purification media by peptide phage display. *Biotechnol. Prog.* **18**, 182–192.

38. Adkins, J.N., Varnum, S.M., Auberry, K.J., et al. (2002) Toward a human blood serum proteome: analysis by multidimensional separation coupled with mass spectrometry. *Mol. Cell. Proteomics.* **1**, 947–955.

39. Chan, K.C., Lucas, D.A., Hise, D., GM et al. (2004) Analysis of the human serum proteome. *Clin. Proteomics J.* **1**, 101–225.

40. Ressom, H.W., Varghese, R.S., Abdel-Hamid, M., et al. (2005) Analysis of mass spectral serum profiles for biomarker selection. *Bioinformatics.* **21**, 4039–4045.

41. Veenstra, T.D., Prieto, D.A., and Conrads, T.P. (2004) Proteomic patterns for early cancer detection. *Drug Discov. Today.* **9**, 889–897.
42. Yu, L.R., Zhou, M., Conrads, T.P., and Veenstra, T.D. (2003) Diagnostic proteomics: serum proteomic patterns for the detection of early stage cancers. *Dis. Markers.* **19**, 209–218.
43. Issaq, H.J., Conrads, T.P., Prieto, D.A., Tirumalai, R., and Veenstra, T.D. (2003) SELDI-TOF MS for diagnostic proteomics. *Anal. Chem.* **75**, 148A–155A.
44. Issaq, H.J., Veenstra, T.D., Conrads, T.P., and Felschow, D. (2002) The SELDI-TOF MS approach to proteomics: protein profiling and biomarker identification. *Biochem. Biophys. Res. Commun.* **292**, 587–592.
45. Rosenblatt, K.P., Bryant-Greenwood, P., Killian, J.K., et al. (2004) Serum proteomics in cancer diagnosis and management. *Ann. Rev. Med.* **55**, 97–112.
46. Diamandis, E.P. (2004) Analysis of serum proteomic patterns for early cancer diagnosis: drawing attention to potential problems. *J. Natl. Cancer Inst.* **96**, 353–356.
47. Koomen, J.M., Li, D., Xiao, L.C., et al. (2005) Direct tandem mass spectrometry reveals limitations in protein profiling experiments for plasma biomarker discovery. *J. Prot. Res.* **4**, 972–981.
48. Semmes, O.J., Feng, Z., Adam, B.L., et al. (2005) Evaluation of serum protein profiling by surface-enhanced laser desorption/ionization time-of-flight mass spectrometry for the detection of prostate cancer: I. Assessment of platform reproducibility. *Clin. Chem.* **51**, 102–112.
49. Baggerly, K.A., Morris, J.S., Wang, J., Gold, D., Xiao, L.C., and Coombes, K.R. (2003) A comprehensive approach to the analysis of matrix-assisted laser desorption/ionization-time of flight proteomics spectra from serum samples. *Proteomics.* **3**, 1667–1672.
50. Baggerly, K.A., Morris, J.S., and Coombes, K.R. (2004) Reproducibility of SELDI-TOF protein patterns in serum: comparing datasets from different experiments. *Bioinformatics.* **20**, 777–785.
51. Baggerly, K.A., Morris, J.S., Edmonson, S.R., and Coombes, K.R. (2005) Signal in noise: evaluating reported reproducibility of serum proteomic tests for ovarian cancer. *J. Natl. Cancer Inst.* **97**, 307–309.
52. Coombes, K.R., Tsavachidis, S., Morris, J.S., Baggerly, K.A., Hung, M.C., and Kuerer, H.M. (2005) Improved peak detection and quantification of mass spectrometry data acquired from surface-enhanced laser desorption and ionization by denoising spectra with the undecimated discrete wavelet transform. *Proteomics.* **5**, 4107–4117.
53. Coombes, K.R. (2005) Analysis of mass spectrometry profiles of the serum proteome. *Clin. Chem.* **51**, 1–2.
54. Coombes, K.R., Fritsche, H.A., Jr., Clarke, C., et al. (2003) Quality control and peak finding for proteomics data collected from nipple aspirate fluid by surface-enhanced laser desorption and ionization. *Clin. Chem.* **49**, 1615–1623.
55. Baggerly, K.A., Edmonson, S.R., Morris, J.S., and Coombes, K.R. (2004) High-resolution serum proteomic patterns for ovarian cancer detection. *Endocr. Relat. Cancer.* **11**, 583–584.
56. Benjamini, Y., Drai, D., Elmer, G., Kafkafi, N., and Golani, I. (2001) Controlling the false discovery rate in behavior genetics research. *Behav. Brain Res.* **125**, 279–284.
57. Morris, J.S., Coombes, K.R., Koomen, J., Baggerly, K.A., and Kobayashi, R. (2005) Feature extraction and quantification for mass spectrometry in biomedical applications using the mean spectrum. *Bioinformatics.* **21**, 1764–1775.
58. Pounds, S. and Cheng, C. (2004) Improving false discovery rate estimation. *Bioinformatics.* **20**, 1737–1745.
59. Pounds, S. and Cheng, C. (2005) Sample size determination for the false discovery rate. *Bioinformatics.* **21**, 4263–4271.

Chapter 15

Analysis of Protein–Protein Interaction Using ProteinChip Array-Based SELDI-TOF Mass Spectrometry

Gitanjali Jayachandran, Jack A. Roth, and Lin Ji

Abstract

Protein–protein interactions are key elements in the assembly of cellular regulatory and signaling protein complexes that integrate and transmit signals and information in controlling and regulating various cellular processes and functions. Many conventional methods of studying protein–protein interaction, such as the immuno-precipitation and immuno-blotting assay and the affinity-column pull-down and chromatographic analysis, are very time-consuming and labor intensive and lack accuracy and sensitivity. We have developed a simple, rapid, and sensitive assay using a ProteinChip array and SELDI-TOF mass spectrometry to analyze protein–protein interactions and map the crucial elements that are directly involved in these interactions. First, a purified "bait" protein or a synthetic peptide of interest is immobilized onto the pre-activated surface of a PS10 or PS20 ProteinChip and the unoccupied surfaces on the chip are protected by application of a layer ethanolamine to prevent them from binding to other non-interactive proteins. Then, the target-containing cellular protein lysate or synthetic peptide containing the predicted amino acid sequence of protein-interaction motif is applied to the protected array with immobilized bait protein/peptide. The nonspecific proteins/peptides are washed off under various stringent conditions and only the proteins specifically interacting with the bait protein/peptide remain on the chip. Last, the captured interacting protein/peptide complexes are then analyzed by SELDI-TOF mass spectrometry and their identities are confirmed by their predicted distinctive masses. This method can be used to unambiguously detect the specific protein–protein interaction of known proteins/peptides, to easily identify potential cellular targets of proteins of interest, and to accurately analyze and map the structural elements of a given protein and its target proteins using synthetic peptides with the predicted potential protein interaction motifs.

Key words: SELDI-TOF-MS, ProteinChip, Peptide interactions, SEND-ID, Fus1, Apaf1

1. Introduction

Signal transduction events in eukaryotic cells involve the reversible assembly of large multi-protein complexes which integrate and transmit the information that controls various cellular processes such as ion fluxes, cytoskeletal rearrangements, gene expression, cell cycle progression, and apoptosis (1, 2). Protein–protein interactions are key elements in the assembly of functional cellular regulatory and signaling protein complexes. Many proteins involved in intracellular signaling contain multiple distinct sequence modules that direct their constitutive and signal-regulated interaction with other proteins in a signaling network (3, 4). These interaction domains can target proteins to a specific subcellular location, provide a means for recognition of protein post-translational modifications such phosphorylation, mediate the formation of multi-protein signaling complexes, maintain functional conformations, and control substrate specificity of enzymes (5, 6). Some of these interactions are mediated by structurally conserved protein domains which recognize specific short peptide motifs such as the well characterized Src-homology SH2 and SH3 domains and the phosphotyrosine-binding PTB domains (7) in protein tyrosine kinases (PTKs). The most recently identified PDZ domains are modular protein-binding domains that can bind to specific recognition sequences at the C-termini of unrelated proteins or dimerize with other PDZ domains or bind to internal peptide motifs to create networks associated with the plasma membrane (8, 9). Hence, deciphering protein–protein interaction cascades is essential for understanding cellular functions of proteins.

Analysis of protein–protein interaction and determination of the precise mechanism and communication among key components in an interactive protein complex such as apoptosome is a very complicated and time-consuming process. The conventional methods currently used to screen potential interacting protein partners, such as the yeast two hybrid assay or phage display systems, are notorious in generating excessive false positives. The Tandem Affinity Purification (TAP) method developed at EMBL (Heidelberg, Germany) has been shown to be a powerful tool for capturing and analyzing cellular proteins potentially interacting with each other under physiological conditions without prior knowledge of the protein composition, activity, or function (10). The TAP assay when combined with mass spectrometry analysis allows the identification of cellular protein targets interacting with a given protein. However, this process is very complex, needs to clone the gene of interest in the TAP-tagged vectors, and requires stringent and lengthy purification efforts (10). Moreover, the TAP system requires validation of targets by means of classical co-immunopurification/immuno-bloting experiments and or gel filtration chromatography. The biochemical approach with FPLC or HPLC for studying protein–protein interaction requires not only expensive equipment, but also highly purified proteins and substantial expertise of the user. To circumvent these operational difficulties, a rapid and sensitive approach is needed for accurately and effectively analyzing protein–protein interaction. Mass spectrometry can serve as a powerful and sensitive tool to attain that goal (11).

We have developed a simple, rapid, and sensitive assay using a ProteinChip array and SELDI-TOF mass spectrometry to analyze protein–protein interactions and map the crucial elements that are directly involved in these interactions. First, a purified "bait" protein or a synthetic peptide of interest is immobilized onto the pre-activated surface of a ProteinChip array and the unoccupied surfaces on the array are protected by application of ethanolamine to prevent them from binding to other non-interactive proteins. Then, the target-containing cellular protein lysate or synthetic peptide containing the predicted amino acid sequence of protein-interaction motif is applied to the protected array with immobilized bait protein/peptide. The nonspecific proteins/peptides are washed off under various stringent conditions and only the proteins specifically interacting with the bait protein/peptide remain on the chip. Last, the captured interacting protein/peptide complexes are then analyzed by SELDI-TOF mass spectrometry and their identities are confirmed by their predicted distinctive masses. This method can be used to unambiguously detect the specific protein–protein interaction of known proteins/peptides, to easily identify potential cellular targets of proteins of interest, and to accurately analyze and map the structural elements of a given protein and its target proteins using synthetic peptides with the predicted potential protein interaction motifs. This simplified assay system with ProteinChip array-based SELDI-TOF-MS platform can also ideally be used in the confirmatory studies following exploratory studies with the yeast two-hybrid system, the TAP-tag assay, and the computer-aided protein interaction modeling (see Note 1). Several databases and bioinformatics tools with a multitude of algorithms, such as BOND (Biomolecular Object Network Databank), PIP (Potential Interactions of Proteins) (12), and PDB (Protein Data Bank), are currently available for the prediction and visualization of interacting protein partners. "Interweaver" is one other versatile program developed by the Institute for InfoComm Research in Singapore that uses BIND, DIP (Database of Interacting proteins) and PDB collectively for generating experimental data. The hit candidates obtained from these databases can be applied in this SELDI-TOF-based protocol to confirm not just direct protein interactions but also the exact interacting motifs of the individual protein partners.

2. Materials

2.1. Protein Chips

Several ProteinChip arrays with chemically modified surfaces (pre-activated) that form covalent bonds with free amine groups are available from Bio-Rad Laboratories (Hercules, CA) (see Note 2). They are developed for diverse biological applications including biomarker discovery, protein profiling, protein–protein interaction studies, peptide mapping for protein ID, immunoassays, and receptor-ligand binding studies. These arrays have eight 2-mm diameter spots (A–H format) that correlate with the spacing of wells in a single column of a standard 96-well microplate and are therefore amenable for high-throughput applications in varied robotics. Here, we describe two types of arrays that are particularly amenable to covalently immobilize biomolecules for subsequent capture of targeted proteins from complex biological samples. The arrays differ in their surface chemistry and hence both should be tested to determine the most suitable array for the application envisioned. Other considerations include lower non-specific binding and higher sensitivity.

1. PS10: The pre-activated surface is composed of carbonyl diimidazole moities to capture amine groups on the protein.
2. PS20: The pre-activated surface is composed of epoxy moities to capture amine groups on the protein.

2.2. Reagents

1. EAM (energy absorbing molecule): CHCA (α-Cyano-4-hydroxycinnamic acid) was purchased from Ciphergen Biosystems (Freemont, CA). These molecules can now be procured from Bio-Rad Laboratories (Hercules, CA).
2. Molecular weight standards (All-in-One Peptide) were purchased from Ciphergen Biosystems (Freemont, CA). They can now be procured from BioRad Laboratories (Hercules, CA).
3. Ethanolamine was purchased from Sigma (St. Louis, MO). Ethanolamine (ETA), also called 2-aminoethanol or mono-ethanolamine (MEA), is an organic chemical compound that is both a primary amine (due to an amino group in its molecule) and a primary alcohol (due to a hydroxyl group).

2.3. Peptide Synthesis

The peptides used in this study were formulated in the Peptide Synthesis Facility at The University of Texas M. D. Anderson Cancer Center.

1. FUS1: KLRRVHKNLIPQGIVKLDHER (2,418 Da).
2. Stearated FUS1: Stearate-KLRRVHKNLIPQGIVKLDHER (2,767 Da).
3. Mutant FUS1: KREHLRKNRPQPGPKPREDHR (2,386 Da).
4. Apaf1: LGILYILQTLE (1,272 Da).

3. Methods

A schematic representation of the overall methodology is depicted in Fig. 1. PS10 arrays are processed as described by Liu and coworkers (13) (see Note 3).

1. Apply 10 μl (1 ng/μl) of bait protein or peptide to individual spots on PS10 chips as bait molecules. The arrays are incubated at room temperature in a humidified chamber for 1 h.

2. Add 10 μl of 1 M ethanolamine onto each spot of the array to block the unoccupied functional surface and incubate at room temperature for 30 min.

3. Wash individual spots three times with (10 μl) of 1× PBS, pH 7.5, containing 0.25% Triton X100 (see Note 4).

4. Apply the second target protein or peptide (10 μl: concentration to be determined empirically) onto the spots containing the immobilized bait protein/peptide and incubated in a humidity chamber for 5–8 h at room temperature or overnight at 4°C.

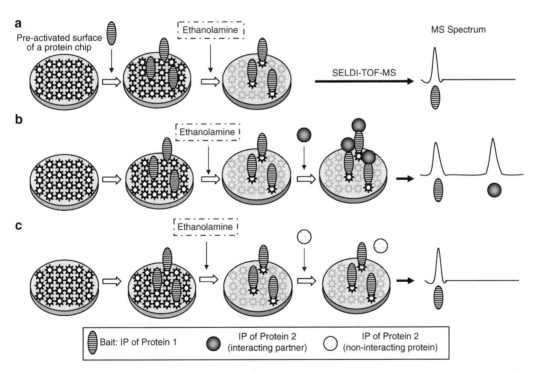

Fig. 1. Schematic presentation of analysis of the protein–protein interaction by an ethanolamine protection on a ProteinChip array with SELDI-TOF-MS. (a) a single peak will be detected in the mass spectrum when a single protein/peptide is immobilized onto the array surface: (b) two peaks will be detected when the target protein/peptide specifically interacts with the bait; and (c) failure of the second peak to show up in the mass spectrum data when the protein/peptide applied subsequently does not interact with the bait protein/peptide.

5. Wash the captured targets on each spot three times in (10 μl) of PBS, pH 7.5, containing 0.25% Triton X100 (see Note 4).

6. Quickly wash each spot two times in distilled water without detergent and air-dry the surface.

7. Add 1 μl of 25× diluted EAM solution (CHCA) onto each spot and air-dry.

8. Acquire mass spectrum using the ProteinChip reader.

9. The molecular mass of the resultant peptide peaks are analyzed and confirmed by externally calibrating the spectra with All-in-One peptide standards with pre-determined molecular weights (Bio-Rad Laboratories, Hercules, CA).

We used this method to study the molecular events in the mitochondria-dependent intrinsic apoptosis pathway mediated by the tumor suppressor FUS1-Apaf-1 protein–protein interaction in lung cancer cells. One of the FUS1-mediated tumor suppressing activities is its ability to induce apoptosis when ectopically expressed by wt-FUS1 gene transfer in FUS1-deficient tumor cells in vitro and in vivo (14, 15). However, the exact molecular mechanism in FUS1-induced apoptosis is unknown. To determine the mechanism of the FUS1-mediated tumor suppression and apoptosis induction, we carried out immuno-precipitations using rabbit anti-FUS1 polyclonal antibodies to pull down FUS1 and its interacting cellular protein partners in crude protein lysates prepared from FUS1-transfected lung cancer cells. We identified one of the potential cellular targets of FUS1 to be the Apaf-1 protein, which has been extensively studied and implicated to be a major component in the apoptosome found in the intrinsic apoptosis pathway (16). We applied this technique to confirm the FUS1-Apaf-1 protein interaction and identify the interacting motifs between two proteins. A computer-aided prediction of the functional motifs in FUS1 and Apaf1 protein sequence revealed a Class I and Class II PDZ protein binding motif in FUS1 and Apaf-1 proteins, respectively. We synthesized peptides containing wild-type or mutated amino acid sequences of the predicted PDZ motifs in both proteins. As demonstrated in Fig. 2, we alternatively used the peptide derived from the PDZ domain of Apaf1 protein or from the PDZ or other functional motifs of FUS1 protein as a bait to test against different peptides derived from both proteins with a wild-type or a mutated amino acid sequence within their PDZ motifs, using PS10 array (see Note 5) by

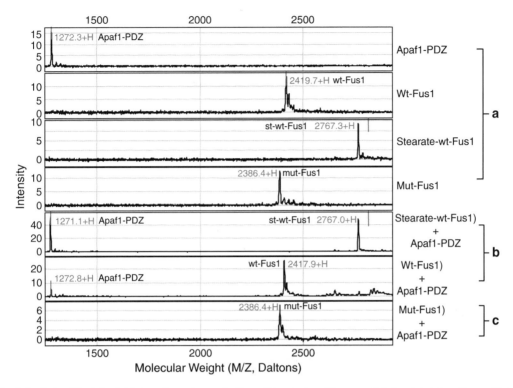

Fig. 2. Analysis of Fus1-Apaf-1 interaction using synthetic peptides designed from predicted protein interaction motifs by an ethanolamine protection assay on a ProteinChip array with SELDI-MS. The specific interaction of the Fus1 PDZ domain with Apaf-1 c-terminal peptide was detected as indicated by accurate mass of each peptide and by comparison with negative mutant and nonspecific control peptides. (a) Panels show individual peaks of various Fus1 or Apaf1 peptides when single peptides were loaded onto proteinchip arrays; (b) two distinctive peaks are detected only when the peptides specifically interact with each other; and (c) when a mutation is introduced in the PDZ binding motif of Fus1, no Apaf1 peptide peak appears in the spectrum.

SELDI-TOF-MS analysis. The specific interaction between the FUS1 and Apaf-1 proteins is clearly demonstrated by the interaction between the peptides derived from wild type amino acid sequences containing the PDZ protein-binding motifs from both proteins. However, no interaction in the non-PDZ sequence containing peptide and the mutated peptides can be detected (Fig. 2). The other attractive feature of this technique is the sensitivity of detection. As demonstrated in Fig. 3, even a very low concentration of peptides and their interactions can be detected. Protein/peptide quantity as low as 10 ng can easily be detected using this method with several modifications and different arrays.

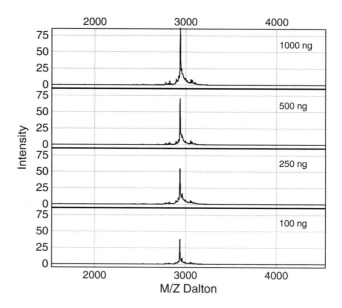

Fig. 3. Quantification of peptide bound to the surface and determination of detection sensitivity on PS10 ProteinChip array by SELDI-TOF-MS. A serial dilution of known concentrations of purified synthetic peptides is loaded onto PS10 chip individually and the amount of peptide is quantified based on the peak intensity.

4. Notes

1. For the characterization of protein-interacting motifs, this technique requires prior knowledge of candidate molecules to be tested and hence cannot function as a screening strategy. It can be useful for validating and confirming the observation and knowledge obtained from screening procedures or database searches.

2. This study describes the use of SELDI-TOF based proteinchip arrays to capture interacting partners after covalent immobilization of bait protein/peptide molecules onto pre-activated surfaces. According to the manufacturer, these arrays are pre-activated with carbonyldiimidazole that have a high affinity toward amino groups. Hence, the amino groups in the peptides applied onto the spots form durable covalent bonds with carbonyldiimidazole already present on the arrays.

3. Akin to other mass spectrometry-based proteomic techniques, crude samples are not amenable to this procedure. Quality of the input protein sample determines the quality of the data generated with it. A high degree of purity in the protein/peptide samples loaded onto arrays is desirable to generate best results and avoid ambiguous and erroneous data. The best results are obtained with pure synthetic peptides. In our hands,

partially purified protein samples did work to a considerable extent, although they did not generate spectacular data. However, it can be used for partially purified over-expressed protein candidates. Liu and coworkers (16) used this elegant technique to determine the interaction between calbindin-D_{28k} and caspase 3 in their study of glucocorticoid-induced apoptosis in osteocytes and osteoblasts. Their study provides a real-world example of application of this technique to decipher meaningful molecular interactions.

4. Stringency of the washes (amount of detergent used in the wash buffer) should be experimentally determined for each project based on the sensitivity the assay and specificity of the binding involved.

5. This procedure can also be accomplished using another proteinchip array with a slightly different chemistry, PS20. It can also be performed using other compatible blocking agents between the two protein-binding steps. As suggested by the manufacturer, Tris–HCl or Glycine (0.1–0.5 M), pH 8.0 can also function as a blocking vehicle.

Acknowledgments

This work was partially supported by grants from NIH/NCI SPORE P50CA070907, RO1CA116322, DOD PROSPECT W81XWH-0710306; and the M. D. Anderson Cancer Center Support Core Grant (CA16672) for using the Peptide Synthesis Facility to synthesize all the peptides used in this study.

References

1. Yaffe, M. B. and Cantley, L. C. (1999) Signal transduction. Grabbing phosphoproteins. *Nature.* **402**, 30–31.
2. Yaffe, M. B. and Smerdon, S. J. (2001) PhosphoSerine/threonine binding domains: you can't pSERious? *Structure.* **9**, R33-R38.
3. Schlessinger, J. (2002) A solid base for assaying protein kinase activity. *Nat. Biotechnol.* **20**, 232–233.
4. Schlessinger, J. and Lemmon, M. A. (2003) SH2 and PTB domains in tyrosine kinase signaling. *Sci. STKE.* **2003**, RE12.
5. Pawson, T. and Nash, P. (2000) Protein-protein interactions define specificity in signal transduction. *Genes Develop.* **14**, 1027–1047.
6. Pawson, T. and Nash, P. (2003) Assembly of cell regulatory systems through protein interaction domains. *Science.* **300**, 445–452.
7. Fanning, A. S. and Anderson, J. M. (1999) Protein modules as organizers of membrane structure. *Curr. Opin. Cell Biol.* **11**, 432–439.
8. Fanning, A. S. and Anderson, J. M. (1998) PDZ domains and the formation of protein networks at the plasma membrane. *Curr. Topics Microbiol. Immunol.* **228**, 209–233.
9. Fanning, A. S. and Anderson, J. M. (1999) Protein modules as organizers of membrane structure. *Curr. Opin. Cell Biol.* **11**, 432–439.
10. Rigaut, G., Shevchenko, A., Rutz, B., Wilm, M., Mann, M., Séraphin B. (1999) A generic protein purification method for protein complex characterization and proteome exploration. *Nat. Biotechnol.* **17**, 1030–1032.
11. Wilkins, M. R., Gasteiger, E., Gooley, A. A., Herbert, B. R., Molloy, M. P., Binz, P. A., Ou, K., Sanchez, J. C., Bairoch, A., Williams, K. L.,

and Hochstrasser, D. F. (1999) High-throughput mass spectrometric discovery of protein post-translational modifications. *J. Mol. Biol.* **289**, 645–657.
12. Jonsson, P. F., Cavanna, T., Zicha, D., and Bates, P. A. (2006) Cluster analysis of networks generated through homology: automatic identification of important protein communities involved in cancer metastasis. *BMC Bioinformatics.* **7**, 2–14.
13. Liu, Y., Porta, A., Peng, X., Gengaro, K., Cunningham, E. B., Li, H., Dominguez, L. A., Bellido, T., and Christakos, S. (2004) Prevention of glucocorticoid-induced apoptosis in osteocytes and osteoblasts by calbindin-D-28k. *J. Bone Miner. Res.* **19**, 479–490.
14. Ito, I., Ji, L., Tanaka, F., Saito, Y., Gopalan, B., Branch, C. D., Xu, K., Atkinson, E. N., Bekele, B. N., Stephens, L. C., Minna, J. D., Roth, J. A., and Ramesh, R. (2004) Liposomal vector mediated delivery of the 3p *FUS1* gene demonstrates potent antitumor activity against human lung cancer in vivo. *Cancer Gene Ther.* **11**, 733–739.
15. Kondo, M., Ji, L., Kamibayashi, C., Tomizawa, Y., Randle, D., Sekido, Y., Yokota, J., Kashuba, V., Zabarovsky, E., Kuzmin, I., Lerman, M., Roth, J., Minna, J. D. (2001) Overexpression of candidate tumor suppressor gene FUS1 isolated from the 3p21.3 homozygous deletion region leads to G1 arrest and growth inhibition of lung cancer cells. *Oncogene.* **20**, 6258–6262.
16. Liu, X., Miller, C. W., Koeffler, P. H., and Berk, A. J. (1992) The p53 activation domain binds the TATA box-binding polypeptide in Holo-TFII-D, and a neighboring p53 domain inhibits transcription. *Cell.* **13**, 3291–3300.

Chapter 16

Quantitation of Amyloid Beta Peptides in CSF by Surface Enhanced MALDI-TOF

Eddie Takahashi, Anita Howe, Ole Vesterqvist, and Zhaosheng Lin

Abstract

Alzheimer's disease is characterized by the deposition of amyloid plaques in the brain. The major components of these plaques are β-amyloid (Aβ) peptides. The CSF concentration of these peptides can therefore provide a valuable biomarker for potentially predicting the state of disease and/or monitoring the efficacy of a drug aiming to inhibit the formation of amyloid plaques. Although the concentration of a given peptide in CSF can easily be measured by ELISA methods, few methods are able to simultaneously observe and distinguish between various peptides of similar yet slightly different amino acid composition. The Surface Enhanced Laser Desorption/Ionization–Time Of Flight mass spectrometry (SELDI-TOF) technology, a platform combining the use of an antibody and MALDI-TOF, can be used to simultaneously detect and quantitate various Aβ peptides with sensitivities in the picomolar range.

Key words: β-Amyloid peptides, Aβ40, Aβ42, Alzheimer's disease, MALDI-TOF, SELDI-TOF, CSF, Quantitation

1. Introduction

The most common form of dementia, Alzheimer's disease (AD) is a progressive neurodegenerative disorder resulting in a gradual loss of memory and major cognitive functions (1). A major characteristic of AD involves the formation of extracellular deposits of amyloid plaques in the brain (2, 3). These plaques have been described as being primarily composed of β-amyloid (Aβ) peptides, which are fragments resulting from a sequential cleavage of a large transmembrane precursor, the amyloid precursor protein (APP), by β- and γ-secretases (4). Due to a lack of specificity in the cleavage performed by the γ-secretase complex, a series of Aβ peptides of various lengths can be formed, all sharing a same N-terminus but with

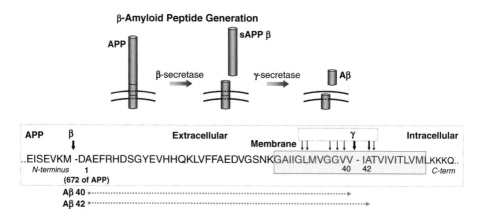

Fig. 1. β-Amyloid peptides result from the cleavage of the amyloid precursor protein (APP) by the β- and γ-secretases. The γ-secretase cleaves primarily 40 residues downstream of the β cleavage site but can also cleave, to a lesser degree, at adjacent sites thereby producing amyloid peptides of various lengths.

a varying number of residues (see Fig. 1). Aβ42 and, to a lesser extent, Aβ40 are considered the most important Aβ fragments in the development of AD (5, 6).

A number of monoclonal antibodies are commercially available to specifically recognize sequence epitopes in Aβ40, Aβ42, or other Aβ peptides, thus allowing an equally significant development of ELISA kits that can readily and quantitatively measure select Aβ peptides in human tissue samples. The technology offered by Surface Enhanced Laser Desorption/Ionization–Time Of Flight mass spectrometry (SELDI-TOF) platform, which combines the unique strengths of immunoassay and MALDI-TOF technologies, allows to essentially perform an immunoassay-type enrichment of peptides binding to an antibody-coated surface prior to quantitation using the mass-based specificity of MALDI-TOF detection. This technique has been extensively used for the identification of novel biomarkers but considerably less as a true quantitative tool owing to the notoriously limited quantitative possibilities of MALDI-TOF mass spectrometry. Typically, MALDI ionization of a given peptide is not reflective of the quantity of the peptide available. Various factors can affect the intensity of a peak signal, ranging from the size and amino-acid content of a peptide, to the type of MALDI matrix and mass spectrometry parameters used. By including an internal standard and controlling certain key aspects of the MALDI-TOF mass spectrometry, we have successfully developed and validated a quantitative SELDI-TOF method for quantitation of multiple Aβ peptides in human cerebrospinal fluid (CSF) (7). A significant advantage of this method is the possibility to quantitate several Aβ peptides simultaneously with high specificity and acceptable sensitivity.

2. Materials

1. Monoclonal antibody 6E10 (Cell Sciences, Canton, MA).
2. Diluent (Standard ELISA diluent, WAKO, Richmond, VA).
3. Aβ peptides (Anaspec, San Jose, CA) are reconstituted in water at 1 mg/mL (see Note 1). Prepare aliquots and freeze at −20°C or below (see Note 2).
4. Internal standard (IS, see Note 3) peptide (Custom order for synthesis of 24 residue peptide DAEFRHDSGYEVH HQKLVFFAEDG, average neutral mass 2,834.0 Da, Anaspec) is reconstituted in water to make 1 mg/mL (see Note 1). Make a subsequent tenfold dilution in diluent to reach a final concentration of 100 µg/mL of IS. Make aliquots and freeze at −20°C (see Note 2).
5. Human cerebrospinal fluid (CSF) (Bioreclamation, Hicksville, NY). Store at −80°C.
6. MALDI matrix: Reconstitute 5 mg α-cyano-4-hydroxycinnamic acid (CHCA, Bio-Rad, Hercules, CA) in 100 µL acetonitrile and 100 µL 1% TFA. Vortex well and sonicate for 15 min. Add 800 µL of 50% acetonitrile, mix and centrifuge at $1,500 \times g$ for 5 min in a microcentrifuge. Store at room temperature (RT), well sealed and protected from light. Good for 30 days from day of preparation.
7. Water (Milli-Q® grade): Unless otherwise specified, all requirements for water throughout this method imply the use of pure water with a resistance of 18 MΩ.
8. Acetonitrile (VWR International, West Chester, PA).
9. Trifluoroacetic acid (TFA, Sigma-Aldrich, St. Louis, MO).
10. Prepare a bovine serum albumin (BSA, Sigma-Aldrich) solution at 2 mg/mL in water.
11. Prepare a 0.5% *n*-octyl glucoside (NOG, Sigma-Aldrich) solution in water.
12. Phosphate-buffered saline (PBS, VWR International).
13. Prepare 1% ammonium hydroxide (VWR International) in water.

3. Methods

These instructions assume the use of a BioRad (Hercules, CA) ProteinChip 4000 Enterprise SELDI-TOF mass spectrometer equipped with PS20 Arrays and Bioprocessors. The method includes two overnight incubations, thus a 3-day experiment can be described as follows: Day 1: antibody coating; day 2: sample loading; and day 3: MALDI-TOF analysis.

3.1. Antibody Coating of Array

1. Dilute the monoclonal antibody in PBS to reach a concentration of 0.25 mg/mL (see Note 4). Pipet 4 µL of diluted antibody onto each spot of a PS-20 array (see Note 5).
2. Place array in a humidity chamber (see Note 6) and incubate overnight at 4°C (see Note 7).

3.2. Sample Antigen Binding

1. Remove array from the humidity chamber.
2. Remove the excess antibody solution by carefully blotting each spot with the corner of a paper tissue, avoiding a direct contact with the spot surface.
3. Immediately (see Note 8) insert the array(s) into a bioprocessor (see Note 9). Use a 96-well plate cover as a protective lid covering the bioprocessor, allowing to minimize cross-contamination and splashes. Briefly centrifuge the bioprocessor (see Note 10).
4. Block non-specific binding sites of the array by adding 30 µL of BSA (2 mg/mL) to each spot. Incubate at RT for 30 min on a shaking platform.
5. Remove the BSA by inserting a papercloth between the bioprocessor and the lid, and briefly centrifuging the bioprocessor upside-down.
6. Wash the arrays by adding to each spot 30 µL of 0.5% *n*-octyl glucoside (NOG). Incubate at RT for 5 min on a shaking platform.
7. Remove NOG by centrifugation and repeat NOG wash a second time.
8. Remove NOG and add 30 µL PBS to each spot. Incubate at RT for 5 min on a shaking platform.
9. Repeat PBS wash a second time.
10. Remove PBS and load 150 µL of CSF sample or calibrator solution to each spot and 5 µL IS solution.
11. Incubate overnight at 4°C on a shaking platform (see Note 11).

3.3. Matrix Addition

1. Remove samples and controls/standards and add 30 µL PBS to each spot. Incubate at RT for 5 min on a shaking platform.
2. Repeat PBS wash two more times.
3. Remove PBS and add 30 µL of water to each spot (see Note 12).
4. Remove water immediately and then remove each array from the bioprocessor.

5. Add immediately (see Note 13) 0.5 µL CHCA matrix to each spot. Let dry for at least 5 min.

6. Add a second layer of 0.5 µL CHCA matrix to each spot. Let dry for at least 5 min.

7. Insert each array into the array holder. Insert the array holder into the SELDI-TOF mass spectrometer.

3.4. Mass Spectrometry Parameters

Use the following SELDI-TOF parameters to perform the detection of Aβ peptides.

1. Mode: Advanced Acquisition.
2. Pulser Ion Extraction Delay: 0.55 µs.
3. Detector Blanking Delay: 10 µs.
4. Mass Range: 0–10,000 Da.
5. Sampling Rate: 800 MHz.
6. Calibration: PCS4000 Instrument Default.
7. Acquisition Method: SELDI Quantitation.
8. Shot Sequences:
    (a) 1 shot at 500 nJ, fire laser, remove from spectrum.
    (b) 50 shots at 500 nJ, fire laser, add to spectrum.
9. Partition: 1 of 10 with 50 shots at each of 21 pixels (total number of shots: 1,050).
10. Ion Source Polarity: Positive ions.
11. Ion Source Voltage: 25 kV.
12. Laser Rate: 20 Hz.
13. Laser Focus Type: Calibrated.
14. Pulser Voltage: 3,414 V.
15. Pulse Width: 435 µs.
16. Steady State: 0.5 s.
17. Detector Bias Voltage: 3,250 V.

3.5. Spectra Analysis

1. For each spectrum, an internal mass calibration can be performed using the IS $((M+H)^+ = 2,835.0$ Da) and Aβ40 $((M+H)^+ = 4,331.9$ Da) peaks (see Note 14).

2. Select the peaks of interest such as the peaks corresponding to the internal standard, Aβ40 (see Note 15) or Aβ42 (4,515.1 Da) in each spectrum and generate an Expression Difference Mapping (EDM) table by clustering only the user selected peaks. An example of a mass spectrum obtained from a human CSF sample is presented in Fig. 2. The results expressed as m/z and peak heights can then be transferred to a Microsoft Excel spreadsheet for further analysis.

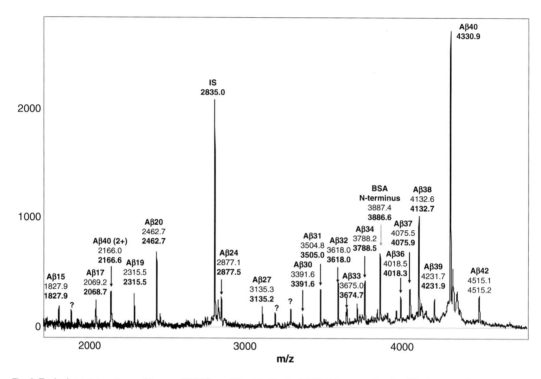

Fig. 2. Typical mass spectrum of human CSF Aβ peptides binding to 6E10. This monoclonal antibody recognizes a sequence epitope located between residues three to eight of the Aβ peptides (EFRHDS). Top m/z values describe the theoretical, expected average $(M+H)^+$ values for each peptide while lower m/z values (*in bold*) depict the experimentally measured values. For Aβ40, an additional doubly charged peak $((M+2H)^{2+})$ can be observed. The spectrum was internally calibrated with the IS and singly charged Aβ40 peaks.

3. Normalize all Aβ peptide values by dividing each peak intensity by the corresponding (i.e. within the same spot) IS peak intensity. Use the normalized values of the various calibrant standards to establish a calibration curve for each Aβ peptide (see Note 16). A typical calibration curve for Aβ42 is presented in Fig. 3. The assay sensitivity should allow the measurement of Aβ peptides at concentrations of 0.2 ng/mL or higher in 100–200 μL of human CSF (see Note 17).

4. Notes

1. Aβ peptide stock preparation: Certain Aβ peptides (especially Aβ42) can exhibit a high level of hydrophobicity, rendering them difficult to reconstitute in water. In such cases, we recommend to initially reconstitute a lyophilized Aβ peptide in 40–50 μL of 1% ammonium hydroxide before completing to 1 mL with Milli-Q® water.

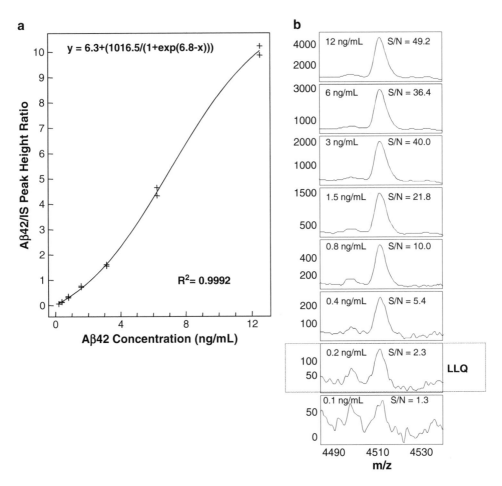

Fig. 3. Example of a calibration curve for Aβ42 and individual mass spectra. (a) The fit model used was based on a four-parameter Boltzmann-type fit with an equation model of $y = (A + ((B-A)/(1 + \exp((C-x)/D))))$ with parameter D equal to 1. (b) The lower limit of quantitation was 0.2 ng/mL for Aβ42.

2. Aβ peptides are remarkably unstable. We have observed that visual precipitates can be formed even while thawing a frozen stock of peptides in water. To reduce the loss occurring from the handling of Aβ peptides in storage or during dilutions of standard calibrators, we have found that proprietary diluent formulations provided by manufacturers of ELISA kits present an efficient alternative to water or PBS for both dilutions and storage purposes. Serum can also be used as an alternative but renders the results more complicated to interpret due to the presence of endogenous Aβ peptides.

3. The use of IS is absolutely essential for any mass spectrometry-based quantitative analysis. The IS should bind to the antibody but ideally should contain a mutation not occurring naturally so that no overlap with an existing peptide is created. In our

case, we have used a custom synthesized peptide corresponding to the first 24 residues of Aβ peptides but with a V24G mutation (average neutral mass of 2,834.0 Da).

4. The antibody concentration of 0.25 mg/mL is deliberately high to maximize the binding onto the arrays as the reaction between the antibody and the epoxide surface is not very efficient. Lower antibody concentrations can be used if necessary.

5. PS-20 arrays provide a reactive surface that can create a covalent bond with a protein such as an antibody. The manufacturer offers other types of arrays, one of which, RS-100, is claimed to be more efficient in producing similar covalent bindings to antibodies. In limited comparison experiments, we have not found any additional benefits using the RS-100 arrays.

6. The humidity chamber is an environment that will minimize the evaporation of the small antibody-containing droplet on an array. A simple emptied pipette-tip box with a wet paper cloth inside can provide such an environment.

7. For fast, same-day experiments, the antibody coating can be performed in two hours. However, as mentioned previously, the reaction between an epoxide and a protein can be very slow and inefficient. For this reason and in order to maximize the sensitivity of the method, we recommend an overnight incubation.

8. After an array has been coated with antibody, we recommend performing all subsequent steps in a timely fashion to avoid drying the array, therefore minimizing any possible denaturation of the antibodies.

9. The sample volume is a critical parameter in increasing the sensitivity of the assay. It is only possible to load a few microliters of CSF sample onto an array, however, we have found that increasing the CSF volume to 150 μL with the use of a bioprocessor allowed us to increase the sensitivity to 0.2 ng/mL of Aβ peptides (about 44 pM).

10. When dealing with a large number of arrays, the use of a multichannel pipette is recommended. A small reservoir can be used to hold the washing solution. For CSF samples, we recommend performing all sample preparations in advance and transferring the samples into a polypropylene 96-well plate using the same plate layout as in the bioprocessor. The samples can then be easily loaded onto the bioprocessor at the appropriate time. Once any reagent or solution is added to the bioprocessor, it is critical that the plate is briefly centrifuged in order to eliminate any air bubbles.

11. As stated previously, the sensitivity of the assay is considerably improved when using higher volumes of sample (i.e. 50–200 μL).

Because only a fraction of the sample is in direct contact with the small surface of the array spot, it is critical to shake the bioprocessor vigorously in order to provide a continuous mixing of the sample content thus exposing more peptides to the covalently bound antibodies.

12. Salts can suppress ionization in MALDI mass spectrometry. The manufacturer therefore recommends briefly rinsing an array with water prior to the MALDI matrix addition. We recommend keeping this practice but have not found any observable difference with or without a water rinse.

13. Peptides and proteins are notoriously susceptible to oxidation when analyzed by MALDI-TOF techniques. Methionine and tryptophan residues are readily oxidized in peptides yielding $(M+H+16)^+$, $(M+H+32)^+$, and $(M+H+48)^+$ peaks. A recently published study identified the source of this oxidation as ambient ozone (8). The study demonstrated that the oxidation was taking place specifically during the drying of the MALDI-TOF spots and hypothesized that part of the peptides may be exposed to the ambient air during the drying process. In the case of Aβ peptides, a methionine residue present at position 35 is susceptible to oxidation (9). All peptides containing this residue have frequently presented a +16 Da peak that is most likely due to ozone oxidation. A simple yet efficient way to minimize the oxidation and the presence of +16 peaks, which can complicate the quantitative interpretation of spectra in SELDI-TOF, consists of applying the MALDI matrix to each spot before the sample spot has been allowed to dry. An exclusive advantage of SELDI-TOF over MALDI-TOF is the binding of an analyte to the surface of a given array, thus refraining the analyte in question to be exposed directly to air. By adding the MALDI matrix before the sample has dried, the exposition to air is minimized, thereby avoiding unnecessary oxidation in a simple yet efficient manner.

14. All MALDI-TOF spectra need to be calibrated. Calibration can be external (i.e. imported from another spectrum) or, ideally, internal using two or more peaks corresponding to peptides of known masses. Internal calibration provides the highest possible mass accuracy. A high level of mass accuracy may not be critical for the quantitation of a given peptide but becomes important when the need to identify an unknown peak arises.

15. A limitation of low resolution MALDI-TOF mass spectrometry is that doubly charged peptides cannot be distinguished from singly charged ones. It is therefore important to verify unaccounted peaks or peaks with considerable difference between their expected and measured masses. A common mistake found in publications of SELDI-TOF studies analyzing Aβ peptides is to misinterpret a spectrum by assigning m/z of

2,166 to Aβ18 (expected $(M+H)^+ = 2,168.3$ Da). The overwhelming presence of Aβ40 yields doubly charged ions with an expected m/z of 2,166.0, a peak easily recognized in internally calibrated spectra.

16. We have found that a linear fit for the calibration curves can generally be used although the best fit was obtained by using a four-parameter Boltzmann-type fit with an equation model of $y = (A + ((B-A)/(1+\exp((C-x)/D))))$ with parameter D equal to 1.

17. When using the conditions and parameters described in this method, Aβ peptides in CSF should be easily detected in the low ng/mL to high pg/mL range when using 100–150 μL of CSF. If the sensitivity is not met, we recommend having the instrument laser source tested by a service engineer to measure the laser energy. On the BioRad ProteinChip 4000 Enterprise SELDI-TOF mass spectrometer, the key parameter reflecting laser energy is referred to as "LEM 1 average." If this value falls below 1,200, the sensitivity will be decreased by several magnitudes.

Acknowledgments

The authors would like to thank Dr. Amanda Bulman for her constructive suggestions and SELDI-related technical support throughout the development of this method.

References

1. Aguero-Torres, H., Fratiglioni, L., and Winblad, B. (1998) *Int J Geriatr Psychiatry* **13**, 755–66.
2. Alzheimer, A., Stelzmann, R. A., Schnitzlein, H. N., and Murtagh, F. R. (1995) *Clin Anat* **8**, 429–31.
3. Tanzi, R. E., and Bertram, L. (2005) *Cell* **120**, 545–55.
4. Pastorino, L., and Lu, K. P. (2006) *Eur J Pharmacol* **545**, 29–38.
5. Blennow, K. (2005) *Expert Rev Mol Diagn* **5**, 661–72.
6. Sjogren, M., Andreasen, N., and Blennow, K. (2003) *Clin Chim Acta* **332**, 1–10.
7. Takahashi, E., Howe, A., Vesterqvist, O., and Linz, Z. (2007) *Proceedings of the 55th ASMS Conference on Mass Spectrometry and Allied Topics, Indianapolis, IN, (June 3–7, 2007)*.
8. Cohen, S. L. (2006) *Anal Chem* **78**, 4352–62.
9. Kadlcik, V., Sicard-Roselli, C., Mattioli, T. A., Kodicek, M., and Houee-Levin, C. (2004) *Free Radic Biol Med* **37**, 881–91.

INDEX

A

Albumin removal 16, 82–85
Alzheimer's disease 110–113, 115, 227, 228
Amniotic fluid 171–196
Amyloid beta 110, 227–236
Analysis
    univariate multivariate 45
    unsupervised multivariate 36, 45, 46
Anion exchange chromatography 23, 24, 28, 30, 52–56, 68, 72, 134–135
Application 2, 7–10, 17, 36, 38, 39, 46, 68, 111, 116, 121, 124–125, 128, 131–140, 160, 179, 181, 182, 192, 204, 213, 219, 220, 225
Arrays 2, 24, 50, 68, 83, 98, 111, 120, 137, 145, 157, 178, 200, 217–225, 229
Assay 1–10, 30, 38, 53, 102, 103, 110, 124, 128, 163, 167, 176, 178, 179, 204, 207, 218, 219, 223, 225, 232, 234
Atrophy 131–140

B

Baseline subtraction 36, 41, 74, 183, 184
Basic fibroblast growth factor (bFGF) 154–156
Biological samples 15, 149, 172–176, 178–181, 188, 194, 201, 220
Biomarkers 2, 35–39, 44, 48–53, 67–78, 95, 98, 110, 111, 113, 120, 124, 133, 143, 154, 156, 166, 171–173, 175, 176, 179, 182, 183, 185–190, 193, 200, 204, 220
Bioprocessor 3, 7, 47, 54, 70, 72–74, 78, 83, 87, 99, 102, 105, 106, 121, 124, 125, 137, 140, 146, 147, 179, 208, 229, 230, 234, 235
Blood-borne parasites 68

C

Calibration
    external 39–40, 74, 116, 127
    internal 39, 40, 65, 74, 90, 91, 116, 235
Calibration standards 39, 40, 89
Cancer 3, 46, 68, 82, 95, 96, 98, 131, 144, 154, 156, 166, 191, 200, 203, 204, 209, 211–213, 220, 222
Candidate biomarker 35–40, 43–45, 47–65

Cell culture media 53
Cerebrospinal fluid (CSF) 98–100, 109–116, 176, 179, 188, 189, 194, 227–236
Cervical lavage 143–151
Cervical mucosa 144, 149, 151
Chromatography 2, 8–10, 12–15, 17, 19–25, 27, 30, 50–53, 63, 68, 72, 98, 99, 102, 105, 106, 120, 132–134, 138, 140, 144, 147, 204, 218
Clinical proteomics 171–196
Concentration range 11–14, 16, 17, 19, 70, 127

D

Data manager software 35–48
Data processing 7, 35–48, 90–95, 209–213
Database searching
    NCBI 62
    SwissProt 62
Depletion 13, 16–18, 24, 25, 28, 68, 82, 84, 85, 95, 203
Discovery 22, 35–37, 47, 51–53, 55, 67–78, 98, 110, 113, 133, 156, 166, 172, 173, 175, 176, 182, 194, 200, 211, 212, 220
Disease 1–3, 38, 45, 62, 67, 68, 76, 81, 82, 95, 97, 98, 110, 113, 143, 154, 156, 167, 172–177, 183, 185–192, 194, 200, 203, 204
Disorders 110, 113, 154, 171–196, 227
Disuse 131–140

E

E. coli lysate 6, 56
Enzymes
    AspN 62, 65
    LysC 65
    Trypsin 62, 65
Experimental design 173, 175

F

Female genital tract 143, 147
Fluid 99, 100, 109, 140, 143, 144, 147, 149–151, 175, 179, 190
Fractionation 12, 13, 15, 21–27, 29, 30, 55, 56, 59, 68–72, 82–86, 95, 104, 106, 111, 120, 133, 135–138, 149–151, 156, 157, 162–164, 204

Index

H

Hepatotoxicity 119
High mass range 37, 41, 88–90, 115
High throughput 2, 25, 144, 173, 175, 199–214, 220

I

Identification 38, 68, 133, 154, 156, 163, 173, 175, 176, 194, 200, 209, 213, 218, 228
Immobilized metal affinity chromatography (IMAC) 19, 20, 25, 26, 51–59, 69, 70, 72, 73, 164, 178
Immunomagnetic separation 213
In-gel digestion 64
Ionization 2, 3, 5, 8–10, 36, 67–78, 82, 119, 132, 172, 228, 235
Isolation 120, 122–124, 153–168

L

Lavage fluid 143–151
Low abundance proteins 16, 17, 21, 29, 68, 70, 145, 149, 213
Low-mass range 37

M

Mass accuracy 40, 173, 175, 235
Matrix assisted laser desorption / ionization TOF (MALDI TOF) 10, 51, 62, 99, 227–236
Matrix attenuation range 37
Muscle tissue 131–140

N

Neat lavage 144
Negative selection marker 53
Nematode. *See* C.elegans
Nephrotoxicity 119
Normal phase arrays 53, 54
Normalization 2, 4–10, 36, 42–43, 75, 90–92, 167, 168, 179, 184, 185, 189, 209

O

Optimization 41, 88, 89, 124, 125, 176, 179–182, 194–196
Osteosarcoma 81–95
Ovarian cancer 3

P

Parasitic diseases 67, 68
Pediatric 81–95
Peptides 2, 17, 35, 49, 83, 98, 113, 149, 158, 183, 218, 227

Phospho-antibody-conjugated magnetic beads 206–207
Phosphoproteins/ peptides 199–214
Plasma 16, 55, 67–78, 81–95, 120, 154–156, 160–163, 166–168, 218
Platelet derived growth factor (PDGF) 154–156
Platelet isolation
    human 161–162
    mouse 162–163
Platelets 153–168
Platform 67–78, 120, 134–136, 171–173, 175, 199–214, 219, 228, 230
Post-translational modification 11, 12, 218
Pre-fractionation 12, 13, 22, 51, 52, 144
Pregnancy 171–196
Profiling 19, 24, 72, 74, 81–95, 97–107, 132, 133, 138, 143–151, 157–158, 164–165, 176, 179–183, 199–214
Protein extraction (TRI reagent) 121, 123
Protein profiling 24, 106, 109–116, 132, 138, 144, 204, 209, 220
Protein-protein Interaction 16, 25, 27, 55, 217–225
ProteinChip 3, 4, 7, 8, 24, 26, 29, 35–48, 51–54, 61, 63, 68–70, 72–74, 83, 87–88, 97–107, 111–116, 120, 121, 124, 134, 135, 145–151, 157, 158, 164, 178, 185, 200, 202, 204, 205, 208–210, 212, 217–225, 229, 236
Proteolytic digestion 49–51, 53, 61, 201, 205
Proteomics 2, 11–30, 49, 53, 57, 68, 81–95, 98, 110, 122, 131–140, 144, 149, 153–168, 171–196, 200–205, 209, 212, 213, 224
Proteominer™ beads 144, 145, 149
Purification 8, 17, 156, 200, 201, 204, 205, 218

Q

Q_HyperD F beads 70, 85, 86
Quantitation 2, 6, 7, 10, 88–90, 194–196, 227–236
Quantitative assay 2, 4–5, 8

R

Reverse phase chromatography 52, 53, 55, 59–60, 63, 64

S

SDS-PAGE 18, 23, 24, 29, 50–53, 59–62, 65, 132, 207
SEND-ID array 202, 208
Serum 2–9, 13, 16, 18–20, 23–26, 28–30, 53, 59, 67–78, 83, 85, 92, 93, 97–100, 109, 111, 116, 128, 137, 140, 149, 154–157, 163, 166, 199–214, 229, 233
Skeletal muscle 131–140
Solid-phase 11–30, 206
Spectral peaks 36, 75

Strategies 11–30, 52, 150, 172, 224
Surface enhanced 190, 227–236

T

Tandem mass spectrometry 50, 53, 61–63
Technology. 12, 36, 68, 97–107, 140, 144, 149,
  150, 157, 172, 173, 201, 204, 206, 209, 212, 213, 228
Theranostics 171–196
Toxicological 119–128

U

Urine 53, 55, 58,
  97–107, 149, 172, 175, 176, 179, 183, 188, 192, 193

V

Variability 5, 9, 10, 36, 38, 99, 100, 104,
  147, 149, 160, 173, 193, 204
Vascular endothelial growth factor (VEGF) 153–156